当代建筑创作理论与创新实践系列

CONTEMPORARY
ARCHITECTURAL
THEORY AND PRACTICE
HIGH-RISE BUILDING
高层建筑

梅洪元 朱莹 主编

黑龙江科学技术出版社

目录 | CONTENTS

理论研究
THEORETICAL RESEARCH

008 新世纪高层建筑发展趋势及其对城市的影响
DEVELOPMENT TREND OF HIGH-RISE BUILDING IN NEW CENTURY AND ITS IMPACT ON CITY
梅洪元 陈剑飞 | Mei Hongyuan Chen Jianfei

011 高层建筑设计的全球趋势研究
GLOBAL TRENDS IN HIGH-RISE DESIGN
安东尼·伍德 菲利浦·欧德菲尔德 | Antony Wood Philip Oldfield

015 超高层建筑与城市空间互动关系研究
THE RESEARCH OF INTERACTION BETWEEN SKYSCRAPERS AND URBAN SPACE
梅洪元 梁静 | Mei Hongyuan Liang Jing

018 城市设计视野中的高层建筑
SKYSCRAPER IN THE VIEW OF URBAN DESIGN
卡瑟利娜·鲍尔斯 黛娜·哈拉萨 | Katharina Borsi Dana Halasa

022 新世纪高层建筑形式表现特征解析
ANALYSIS OF THE FORM AND PERFORMANCE OF HIGH-RISE BUILDINGS IN THE NEW CENTURY
梅洪元 李少琨 | Mei Hongyuan Li Shaokun

026 从正交四方到随意变换——设计创新在高层建筑形式中的角色
FROM THE ORTHOGONAL TO THE IRREGULAR: THE ROLE OF INNOVATION IN FORM OF HIGH-RISE BUILDINGS
菲利浦·欧德菲尔德 安东尼·伍德 | Philip Oldfield Antony Wood

032 高层建筑参数化设计
PARAMETRIC SKYSCRAPERS
陈寿恒 | Chen Shouheng

037 "少即是多"理论的实践新解
NEW PRACTICAL INTERPRETATION OF "LESS^N IS MORE"
山扬·李沧 安德烈斯·阿利亚斯 马德里 何塞·拉蒙·特拉莫耶雷斯 | Sangyup Lee Andres Arias Madrid Jose Ramon Tramoyeres

041 基于规划视角的超高层建筑思考
THINKING OF THE SUPER HIGH-RISE BUILDINGS FROM THE URBAN PLANNING VIEW
吴婷婷 王世福 邓昭华 | Wu Tingting Wang Shifu Deng Zhaohua

044 深圳高层建筑空间造型实态调研
FIELD RESEARCH ON THE SPACE FORM OF SHENZHEN'S HIGH-RISE BUILDINGS
覃力 刘原 | Qin Li Liu Yuan

设计作品
DESIGN WORKS

052 英国碎片大厦
THE SHARD (LONDON BRIDGE TOWER), UK
伦佐·皮亚诺建筑工作室 | Renzo Piano Building Workshop

058 西班牙费拉Porta Fira双子塔
TORRES PORTA FIRA, SPAIN
伊东丰雄 | Toyo Ito

062 杭州华联UDG时代广场
HANGZHOU HUALIAN UDG TIME SQUARE
冯·格康，玛格及合伙人建筑师事务所 | gmp

066 巴西圣保罗无限大楼
SÃO PAULO INFINITY TOWER, BRAZIL
KPF建筑师事务所 | Kohn Pedersen Fox Associates Pc

072 阿拉伯联合酋长国阿布扎比投资管理局总部大厦
ABU DHABI INVESTMENT AUTHORITY HEADQUARTERS, UAE
KPF建筑师事务所 | Kohn Pedersen Fox Associates PC

078 韩国三星瑞草大厦
SAMSUNG SEOCHO, KOREA
KPF建筑师事务所 | Kohn Pedersen Fox Associates PC

084 英国尖塔
THE PINNACLE, ENGLAND
KPF建筑师事务所 | Kohn Pedersen Fox Architects PC

088 韩国东北亚贸易大厦
NORTHEAST ASIA TRADE TOWER, KOREA
KPF建筑师事务所 | Kohn Pedersen Fox Architects PC

092 韩国松岛国际城住宅
SONGDO INTERNATIONAL CITY, SOUTH KOREA
KPF建筑师事务所 | Kohn Pedersen Fox Architects PC

096 香港理工大学赛马会创新楼
JOCKEY CLUB INNOVATION TOWER, HONG KONG POLYTECHNIC UNIVERSITY
扎哈·哈迪德建筑师事务所 | Zaha Hadid Architects

106 阿拉伯联合酋长国The Opus办公大楼
THE OPUS OFFICE BUILDINGS, UAE
扎哈·哈迪德建筑师事务所 | Zaha Hadid Architects

110 马来西亚黎明之塔
SUNRISE TOWER, MALAYSIA
扎哈·哈迪德建筑师事务所 | Zaha Hadid Architects

116	罗马尼亚多罗班蒂大厦 DOROBANTI TOWER, ROMANIA 扎哈·哈迪德建筑师事务所 | Zaha Hadid Architects	
122	埃及尼罗塔 NILE TOWER, EGYPT 扎哈·哈迪德建筑师事务所 | Zaha Hadid Architects	
128	新加坡花拉阁 FARRER COURT, SINGAPORE 扎哈·哈迪德建筑师事务所 | Zaha Hadid Architects	
134	广州国际金融中心 GUANGZHOU INTERNATIONAL FINANCE CENTER 威尔金森·艾尔建筑事务所 | Wilkinson Eyre Architects	
140	广州东塔 GUANGZHOU EAST TOWER 威尔金森·艾尔建筑事务所 | Wilkinson Eyre Architects	
144	沙特阿拉伯Tadawul证券交易大楼 THE TADAWUL STOCK EXCHANGE TOWER, KINGDOM OF SAUDI ARABIA 威尔金森·艾尔建筑事务所 | Wilkinson Eyre Architects	
150	阿拉伯联合酋长国55°旋转塔 55 DEGREES ROTATING TOWER, UAE 威尔金森·艾尔建筑事务所 | Wilkinson Eyre Architects	
154	深圳证券交易所新总部大楼 SHENZHEN STOCK EXCHANGE HEADQUARTERS 大都会建筑事务所 | Office for Metropolitan Architecture（OMA）	
162	美国111第一大街 111 FIRST STREET, USA 大都会建筑事务所 | Office for Metropolitan Architecture（OMA）	
166	法国灯塔高楼 LA TOUR PHARE, FRANCE 大都会建筑事务所 | Office for Metropolitan Architecture（OMA）	
170	墨西哥Bicentenario塔 TORRE BICENTENARIO, MEXICO 大都会建筑事务所 | Office for Metropolitan Architecture（OMA）	
176	天津中钢国际广场 TIANJIN SINOSTEEL INTERNATIONAL PLAZA MAD建筑事务所 | MAD Ltd.	
182	加拿大多伦多梦露大厦 TORONTO ABSOLUTE TOWERS, CANADA MAD建筑事务所 | MAD Ltd.	

186	意大利商品交易会公司总部大楼 TRADE FAIR CORPORATE HEADQUARTERS, ITALY 意大利IaN*建筑设计 ｜ IaN*	
192	韩国青罗城市大厦 CHEONGNA CITY TOWER, KOREA 意大利IaN*建筑设计 ｜ IaN*	
196	丹麦巨拱 CPH ARCH, DK 丹麦3XN建筑事务所 ｜ 3XN	
202	荷兰奈梅亨商务及创新中心 52° NIJMEGEN BUSINESS AND INNOVATION CENTRE FIFTY TWO DEGREES, THE NETHERLANDS Mecanoo建筑事务所 ｜ Mecanoo	
208	美国赫斯特大厦 HEARST TOWER, USA 福斯特及合伙人事务所 ｜ Foster + Partners	
214	美国世界贸易中心二号塔 TOWER TWO ON THE SITE OF THE WORLD TRADE CENTRE, USA 福斯特及合伙人事务所 ｜ Foster + Partners	
220	阿拉伯联合酋长国阿布扎比资本中心塔 ABU DHABI CAPITAL GATE, UNITED ARAB EMIRATES RMJM建筑设计集团 ｜ RMJM Architects	
224	俄罗斯圣彼得堡奥克塔摩天楼 ST. PETERBURGH OKHTA TOWER & CENTRE, RUSSIA RMJM建筑设计集团 ｜ RMJM Architects	
230	土耳其Atasehir Varyap 项目 ATASEHIR VARYAP PROJECT, TURKEY RMJM建筑设计集团 ｜ RMJM Architects	
236	南京河西新城苏宁广场 NANJING SUNING WEST RIVER CITY PLAZA 凯达环球 ｜ Aedas Ltd.	
240	阿拉伯联合酋长国帝国大厦 EMPIRE TOWE, UNITED ARAB EMIRATES 凯达环球 ｜ Aedas Ltd.	
244	阿拉伯联合酋长国迪拜Pentominium 大楼 DUBAI PENTOMINIUM TOWER, UNITED ARAB EMIRATES 凯达环球 ｜ Aedas Ltd.	

理论研究
THEORETICAL RESEARCH

008 新世纪高层建筑发展趋势及其对城市的影响
DEVELOPMENT TREND OF HIGH-RISE BUILDING IN NEW CENTURY AND ITS IMPACT ON CITY

011 高层建筑设计的全球趋势研究
GLOBAL TRENDS IN HIGH-RISE DESIGN

015 超高层建筑与城市空间互动关系研究
THE RESEARCH OF INTERACTION BETWEEN SKYSCRAPERS AND URBAN SPACE

018 城市设计视野中的高层建筑
SKYSCRAPER IN THE VIEW OF URBAN DESIGN

022 新世纪高层建筑形式表现特征解析
ANALYSIS OF THE FORM AND PERFORMANCE OF HIGH-RISE BUILDINGS IN THE NEW CENTURY

026 从正交四方到随意变换——设计创新在高层建筑形式中的角色
FROM THE ORTHOGONAL TO THE IRREGULAR: THE ROLE OF INNOVATION IN FORM OF HIGH-RISE BUILDINGS

032 高层建筑参数化设计
PARAMETRIC SKYSCRAPERS

037 "少即是多"理论的实践新解
NEW PRACTICAL INTERPRETATION OF "LESS^N IS MORE"

041 基于规划视角的超高层建筑思考
THINKING OF THE SUPER HIGH-RISE BUILDINGS FROM THE URBAN PLANNING VIEW

044 深圳高层建筑空间造型实态调研
FIELD RESEARCH ON THE SPACE FORM OF SHENZHEN'S HIGH-RISE BUILDINGS

DEVELOPMENT TREND OF HIGH-RISE BUILDING IN NEW CENTURY AND ITS IMPACT ON CITY

新世纪高层建筑发展趋势及其对城市的影响

梅洪元　陈剑飞　|　Mei Hongyuan　Chen Jianfei

历经百年沧桑的高层建筑，近20年来在中国蓬勃发展、成就非凡。对我国城市环境和社会经济发展起到了极大的促进作用。随着20世纪的悄然离去，世界范围的社会经济结构调整方兴未艾，我国与广大第三世界国家继续携手向城市化进军。在特定的时代背景下，反思我国高层建筑的建设成就，不难看到其背后令人担忧的现实——高层建筑创作理论的匮乏和实践领域的"拿来主义"。尤其是在全球范围内关注生态问题和可持续发展的今天，我们更应客观地分析和评价高层建筑的创作和实践，使其走向健康发展之路。

一、高层建筑自身发展趋势

高层建筑是现代城市发展的产物，也是人类社会需求多样化、聚居环境高密度化的必然结果。随着工业社会向信息社会的转变，社会一体化发展、产业结构的改变，使得当代高层建筑更频繁、高效地介入到社会动态的循环系统中去，建筑规模越来越大，功能也渐趋复杂；同时，人们由于价值观、思维方式、社会心态等深层机制随社会发展而变化，对高层建筑的功能提出了更进一步的要求，从而推动其发展演变。审视新世纪我国相继涌现的高层建筑，无论是创作观念，还是建筑本体，都呈现出新的特点和趋势。

1. 技术表现综合化

信息时代，社会多学科的互相交融与多技术系统的综合集成构成了推动高层建筑发展的整合力量，使得高层建筑以更深、更广、更直观和更具综合性的方式，拓展功能内涵、空间模式和审美形态，从而增加新的功能维度、空间维度和审美维度。尤其是以现代结构技术、轻质高强建筑材料、抗震和防风等抗灾减灾技术的快速发展作为技术支持与现实条件，使得建筑高度不断攀升。有人预测，随着经济实力不断增强，21世纪的亚洲将会成为世界高层建筑发展中心和高度记录竞争的热点地区。

然而建筑高度的迅猛发展也暴露出许多仍需解决的技术问题和负面影响，包括建筑需要承受更强烈的地震力以及风荷载、造成城市某一地区过分拥挤、增加火灾危险性……因此我们对于人类生存与生活空间向高空拓展的探索，应当采取审慎的态度，确定适宜的高层建筑技术发展战略，根据不同城市、地域的具体情况分别对待。注重技术表现的综合化与真实性，从技术视角对高层建筑创作理念进行深层研究，充分发挥技术对于人类文明进步的促进作用，避免因技术表现上的盲目、浮躁而导致设计水平低下。

多元综合已成为当今高层建筑技术表现的一大趋势，我们倡导对于创作中的技术理念在更深层次上的整合，使高层建筑更加能动地发挥其职能和功效，产生更大的经济效益，从而创造高质量、高情感、高和谐的居住环境。

2. 创作观念多元化

在高层建筑创作实践中，一些勇于探索的建筑师以其生机勃勃的创作观念和创新精神，充分把握技术发展给高层建筑的功能、空间、形式带来的新变化和提供的丰富可能性，积极融会当今世界科技发展的最新成果，并创造性地加以利用；注重高层建筑与自然生态的协调，维护环境的生态平衡，提高能源、资源的利用效益；注重高层建筑与城市文化的融合，并与具体经济条件、物质条件等地域基质相结合，在充分应用现代技术的基础上，发挥地区文化的特色与建筑师的创造才能，创造出许多个性化建筑作品。这些作品或者推崇商业化与俚俗化，追求含混与复杂；或运用极端逻辑性和高度夸张的手法，追求标新立异；或者突出现代技术、表现现代材料的精美，以突出纯净的建筑形体表达丰富的内涵；或者注重环境，用现代手法表达对地域、文化的关注。应该看到，文化的传承和技术进步使得人类探索形势和能源的可选择性加大，技术的表现手段也趋于多样化，同时由于不同地区的客观建设条件不同，经济、技术、文化发展的不平衡，必将促使建筑师进行多元化的探索。

但是同时我们也看到，中国建筑正又一次面对西方的全方位冲击，各种思潮和理论被大量引进，外国建筑师各种风格的作品在中国不断出现，尤其是处于转型期的高层建筑，在外部巨大冲击和内蕴的共同作用下，一方面呈现出数量的不断增长，另一方面也出现许多问题，比如盗版的"KPF"形式主义，"炒作"与"创作"之间的困惑、盲目的标新立异等等。对于现阶段中国建筑界的困惑与混乱，我们应当将其视为一个必经的阶段和过程，要通过建筑师的共同努力，尽快走出彷徨，迈向成熟。"面对广阔的创作天地要抛开表面的装饰性的浮躁解读，不应刻意追求所谓创造标志性的愉悦。"[1]中国目前正处于大建设时期，也是文化转型时期，建筑师只有保持平和的心态，才能回归建筑本原。建造高层建筑的目的是为了满足一定的功能要求，而高层建筑功能空间综合化的发展趋势要求我们采取相应综合化的设计观念。因此在创作中必须本着系统的观点，把握形式与功能、建筑与环境的关系，创造出整体功能更加广泛和优越的高层建筑。

3. 建筑形象个性化

高层建筑作为城市生活的重要时空坐标，往往以其宏伟的尺度和巨大的体量给观者以强烈的视觉感受，同时也决定和影响着其所在城市区域的艺术风格和美学价值。建立高层建筑自身的形式和结构之间的协调关系是其形象创作的基础要素，同时恰当的结构体系和细部处理更可以激发建筑师的想象力和创作灵感，创造出富有表现力和时代特质的新形式。

中国有着悠久的历史和独特的文化传统，但快速的经济发展和城市化使许多地域文化传统正在消失。尤其是一些高层建筑形象设计更是违背了中国传统文化，丝毫不考虑城市肌理、尺度的限制，不注重建筑的经济性和技术表现的真实，片面追求高层建筑的标新立异，过于关注令人兴奋的视觉刺激。这样的建筑作品也许可以迎合大众一时的心理需求，但终究无法通过时间的考验，因为高层建筑的真正魅力并不在于其炫目的外表，而应在于其深刻的文化内涵和内在的逻辑性。建筑是一定时期和地域文化的缩影，高层建筑的发展与其文化背景也是相应的。每个城市都有自己的独特风貌，这些地域性因素是高层建筑形式创作的重要依据，认真研究其所在城市的建筑特征和地方风格并加以提炼升华，结合当代先进技术，融入高层建筑语汇之中，这样创造出来的高层建筑才能被称作文化。

4. 近地空间城市化

近年来，"城市及高层建筑的发展呈现出立体化、集约化、复合化的共同趋势"。[2]因此高层建筑近地空间设计越来越重视与基地范围外的城市空间的结合，逐渐趋于向社会开放，与相邻建筑外部空间的界限逐步消除，形成连续通畅的城市公共空间。高层建筑近地空间城市化可以缓解高层对城市空间的压力，为市民提供生活与交往的场所；同时又使底部的商业设施得以共用，把高层建筑与城市功能有机结合，从而发挥更大的整体效益。

高层建筑占据有限的土地，空间组织模式紧凑、高效，但其与外界的交往却是大量性的。如何处理好其内外功能的交叉与协调，保证建筑与外部环境之间交流的顺畅尤为重要。高层建筑近地空间一方面通过与城市交通网络的连接，使建筑自身乃至城市交通得到快速有序地集散，减轻城市交通负荷；另一方面通过与城市公共空间的结合，实现了高层建筑与城市环境的交流，在保留了地面的生态环境的前提下，将城市区域环境加以整合，从而建立完整的城市空间秩序。对于高层建筑来说，通过底部空间与街道、广场、庭院、踏步相结合，相互穿插、相互渗透，并与商场、餐馆等服务设施密切配合，实现了建筑空间与城市的有机串联；对于城市而言，高层建筑底部空间的开放，更加充分地体现城市对人的尊重关怀，丰富城市生活，改变城市概念，增加了城市活力，并使高层建筑从形式与内容的双重意义上，真正成为现代城市的"主角"。

二、高层建筑对城市的深层影响

高层建筑的产生是城市经济增长和土地资源紧张的必然结果，而高层建筑的急速发展又反过来对城市生态结构和城市文化产生威胁。我们回顾二者的发展演变历程可以看出它们相同的进化趋势以及互动关系。尤其是近几年来，高层建筑因其巨大的体量促使建筑空间容量呈几何级数增加，这对城市发展带来的冲击可想而知，给城市空间带来的压力也是空前的。因此，我们必须格外关注高层建筑与城市的深层关联，正确理解高层建筑对城市的影响与作用。

1. 导致城市空间结构均质化

高层建筑是城市空间体系的重要组成部分，随着城市空间形态的内在结构渐趋复杂化和多样性。在纵横交织的空间网络中，高层建筑由于突出的形体特征和超大尺度的空间容量，成为城市的标志性环节，可以帮助城市人群建立起清晰的空间认知意向和明确的方位感。高层建筑产生之初，由于经济原因大多集中布置于城市地价昂贵的中心区，多为单一功能、数量不多、密度不大，往往成为居于主导地位的城市地标，也是制高点，城市空间结构清晰而丰富。

今天，高层建筑的领域已拓展到医院、住宅、学校等建筑类型。建筑空间也从传统的功能单一性中解脱出来，朝着集多种功能为一体的综合化发展。而且在经历了一段无序的发展之后，其盲目建设与缺乏规划，损坏了城市原有的空间肌理，杂乱无章的天际线形成视觉污染，高层化的城市千篇一律，缺乏个性与地域性，城市空间结构也由此变得均质而单一，人们在城市中失去了位置坐标。

高层建筑的布局与城市总体发展方向密切相关，综合考虑城市的三维空间格局、城市天际轮廓线塑造以及基础设施支持系统和实际情况，运用定量和定性的分析方法，合理确定高层建筑发展区域，才能使其成为城市结构中的积极因素。无论是建筑师、业主还是城市管理者必须找到经济利益、城市环境、建筑单体之间的平衡点，使高层建筑的建设实现总体规划，有序发展，既突出特色，又融于环境，保持适宜的建筑密度和丰富的群体形态。一个整体有序的城市结构的建立，应该是城市空间形态、文化形态、视觉秩序等多方面的集合，具有标志性的高层建筑作为城市空间的主导因素，只有与城市空间环境达到良好的匹配与契合时，才能充分发挥效能，有效地传播文化，提升其美学价值。

2. 促进城市交通网络立体化

建立完善合理的高层建筑交通体系，对提高土地经济效益意义重大，并可改善城市的空间结构和社会结构。高层建筑与城市交通网络的立体化连接，有效地缓解了地面交通的压力，提高了运营效率，使行人活动不再局限于常规的人车共行街道，从而减少相互干扰；同时创造全天候的步行环境，抵御气候的不利影响。

高层建筑功能复杂、人流车流量大、出入口众多，其交通量占有城市交通量的相当比例。为了使建筑内的人流、车流迅速方便地疏散，在高层建筑基地内往往设置专门场地和设施，用于交通流的集散、转换、组合、分配以及车辆存放，与城市道路和各种交通枢纽形成复杂的组织方式。

传统意义上，高层建筑与城市交通网络的连接主要是通过步行系统将内部交通纳入城市交通网络。在步行系统中，除常规街道层步道系外，天桥和地下空间所组成的城市非地面步道系统起到了非常重要的补充作用。它们一方面联系着高层建筑及其周围其他建筑的交通厅、中庭以及外部空间，另一方面又与地铁站、汽车站、停车库等城市交通的起始点相连，共同形成有机联系的整体交通网络。此外，随着城市容量的日益扩大，使得高层建筑向地上、地下综合性地发展城市空间成为必然趋势。高层建筑根据不同需求，通过底部结构、主体结构、尽端结构分别与地面、周围城市环境、空中三者之间形成立体交叉网络，大大改善高层建筑的可达性，为高层建筑系统的高效运行提供保障，也为城市创造了一种更合理的聚居结构模式。

3. 加剧城市环境系统地域化

城市环境作为人类生态系统的组成部分，随着社会的发展而不断变化。在当代世界范围内，尤其是发展中国家的高速发展过程，打乱了城市发展秩序，给环境造成负面影响。高层建筑及其建成环境对城市物理环境及城市机能等要素影响较大，其中"高层风"是最主要的问题之一。随着科技的发展，因风荷载引起高楼振动是可以控制的，但由于高层建筑密集而产生的对周围环境的风流影响则较难控制。

在北方寒冷地区，一方面由于高层建筑的巨大体量，在日照作用下向其底部空间投下大片阴影，使落影区内的建筑、广场和道路终日笼罩在寒冷、潮湿、阴暗之中，给人们的生活和工作带来危害；另一方面高层风不仅严重影响步道层的行人活动，而且对建筑物本身使用安全及管理造成威胁，导致高层区域城市生态环境极其恶劣。而在南方，城市大量密集的高层建筑加剧了城市"热岛效应"，导致区域内气温居高不下，严重影响人们的正常生活；持续高温反过来又使建筑能耗增加，形成恶性循环。在我国高层建筑发展已成必然之势，但它对环境的负面影响也日益显现。面对人们对建筑空间的新需求、城市对建筑形式的新协调、资源对建筑热工的新控制、生态气候对建筑形制的新制约等等。如何有效调控高层建筑自身肌理，最大程度地减小其对环境的负面影响，使我们的建成环境与生物圈的生态系统融为一体，是摆在当代建筑师面前的重要课题。

4. 趋向城市文化内涵混沌化

建筑作为容纳人类活动的物质环境，能否成为人们向往的场所，很大程度上取决于它对于人们功能层面需求的满足程度。然而就高层建筑而言，人们寄予它的精神需求更多也更为强烈。不同历史时期的高层建筑集中反映各不相同的文化内涵，体现出不同城市的文化观念。纵观高层建筑的发展史，建筑师通过不懈的探索创造出一批优秀的建筑作品。尤其是国外不少成功的高层建筑重视文化内涵的发掘，注重将建筑功能与新结构、新材料相结合，从环境、功能、空间、造型、构图等方面塑造建筑个性，从而提升城市文化品位。

与国外的高层建筑发展历程相比，我国目前仍处于发展过程中，尤其是一些中小型城市由于种种原因陷入误区，表现在高层建筑创作上，往往出现两种极端现象：一些建筑师在创作中片面关注业主、大众的审美取向和心理需求，追求新、奇、特的变异形体，导致建筑形象缺乏内在逻辑；还有一些人放弃建筑创作的创新追求，简单抄袭和模仿国外建成作品，导致建筑缺乏地域文化特征和城市面貌的千篇一律。"平庸的城市""平庸的建筑"与日俱增，地域文化的特色逐渐消失，高层建筑在走过其辉煌的巅峰状态之后，陷入到深刻的危机当中。事实上，任何高层建筑创作都无法摆脱城市文化的束缚，同时也会对城市文化产生深远影响。我们在认识到高层建筑与城市在空间上的互动关系之后，必须就它的美学问题和文化现象加以探讨，从深层次考察二者在城市文化发展过程中的相互关系，努力探求高层建筑与城市文化的整合关系。

中国建筑业正面临更大的发展机遇，人口及城市发展与用地之间的矛盾使高层建筑的发展成为必然。面对高层建筑与城市发展已经出现的问题，我们必须重新审视自己的创作观念。理性的思考要求我们不能把简单的问题复杂化。虽然人们可以从不同的角度去预测，但本世纪的建筑问题仍然是如何提高人们的居住环境质量问题；同时，我们更不能把复杂的问题简单化，技术层面的结构、材料、节能、生态等问题仍需我们付出更大的努力。立足于城市及其文化发展的高层建筑是时代的选择，从传统文化中汲取精华迎接挑战是建筑师的职责。

参考文献

[1] 庄惟敏. 几个观点、几种状态、几点呼吁——青年建筑师论坛随笔 [J]. 建筑学报，2004（01）：68-69.

[2] 张宇. CBD 现象的启示与高层建筑的近地空间 [J]. 新建筑，2002（02）：44-45.

作者简介

梅洪元　全国工程勘察设计大师
　　　　哈尔滨工业大学建筑学院院长、教授、博士生导师
　　　　哈尔滨工业大学建筑设计研究院院长、总建筑师
　　　　《城市建筑》主编

陈剑飞　哈尔滨工业大学建筑设计研究院副院长、副总建筑师、教授、博士生导师
　　　　《城市建筑》副主编

GLOBAL TRENDS IN HIGH-RISE DESIGN
高层建筑设计的全球趋势研究

安东尼·伍德　菲利浦·欧德菲尔德　| Antony Wood　Philip Oldfield
李靳　译
汤岳　校审

一、全球范围内的高速发展趋势

毋庸置疑，我们正处于前所未有的高层建筑急剧发展期，这种发展具有全球性规模，从莫斯科到中东，从上海到旧金山，越来越密的城市，越来越高的建筑不断涌现。即使与摩天楼建造的黄金时代——20世纪初芝加哥或装饰艺术运动中的纽约相比，我们也很可能正经历高层建筑最高水平的发展期，而且这种发展是全球范围的。剖析其原因，可能会让人出乎意料。

原因之一：地价

地价，一直都是驱使高层建筑发展的因素。不过，越来越多的城市，特别是在美国和英国这样的国家，通过在"商业-零售"为主导的中心商业区穿插"居住-休闲"功能，实现城市中心的复苏。这些相对较新的业态催动城市中心地价的提高，也使建筑高度成为兑现投资回报的必要因素。

原因之二：全球性地标

超高层建筑的建造往往不仅是为追求商业上的回报率，相反很多人相信建筑在超过一定高度后，经济效益并不能像建筑形式那样"节节高"。创造一座凌驾于城市之上的建筑地标，一直都是超高层建筑的建造初衷。当今社会高层建筑成为衡量一座城市在全球范围内重要性的标志，因此各大城市争先创造具有全球品牌认知度的天际线。这一变化，是将经营上的考虑变为城市（甚至是政府）的野心，而这种野心就体现在"世界最高"称号的争夺上。历史上我们有克莱斯勒大厦(Chrysler Building)或西尔斯大厦(Sears Tower)，现在我们有台北101、阿联酋迪拜塔(Burj Dubai)、俄罗斯塔(Russia Tower)和上海环球金融中心(Shanghai World Financial Centre)。这些建筑本身就肩负着在世界舞台上"推销"所在城市的重任，同时也彰显地域性内涵。

原因之三：可持续发展

密度更高、更浓缩的城市，目前被视为更利于可持续发展的生活模式的建立——通过减少城市向郊区的扩张、合理配置交通和基础设施网络，最终减少能量的消耗和有害气体的排放。当然，高层建筑是创造高密度城市的关键一环，它能以最小的占地面积承载更多的人工作、生活。另外，每栋高层建筑项目在经济和技术上的高投入，为可持续理念和生态技术的实践提供了机会。而这些实践对一些小型建筑项目同样具有指导意义。

原因之四：世界贸易大楼的倒塌

世界贸易大楼的倒塌也许是过去半个世纪中发生的最具影响力的事件，它使我们产生了疑问："在后'911'时代我们是否应继续建造高层建筑？"在这7年中，若从高层建筑不断被提议和建造数量来看，答案是肯定的。这一事件曾导致世人对高层建筑的深刻反思，却促进了更全面的设计、更安全的建筑及更好的城市中心产生。政府、城市管理者、金融家、开发商也更多地体会到这场全球范围内自我反思的益处。

自西尔斯大厦为美国赢得"世界最高"的称号以来，高层建筑在过去几十年间发生了很多变化。目前更多的高层建筑集中在亚洲，而不是北美。在2007年竣工的10大超高层建筑中，4栋在中东、4栋在亚洲、1栋在北美、1栋在欧洲。20世纪80年代以前，世界最高的建筑一般会出现在北美，钢结构为主体，功能是办公建筑。今天，这种概念几乎被完全推翻——设计、在建的世界最高建筑均位于亚洲和中东，混凝土建造且功能主要为居住，这也正是在建的"世界最高"的阿联酋迪拜塔的真实写照。单从高度来看，2009年即将完工的迪拜塔将超过800 m，比现今世界最高的台北101大厦还高300 m。

近期芝加哥"高层建筑与城市环境协会(CTBUH)"进行了关于研究"2020年20幢最高建筑"的研究（图1）。此研究是基于建成、已建、在建或"真实的项目计划"展开的（所谓"真实的项目计划"，是指开发方和设计团队正进行的设计项目，且深入程度已超过概念设计的阶段）。研究结果再次证明现今高层建筑的实践活动已离开北美，20栋建筑中的9栋会在亚洲、8栋在中东、2栋在北美、1栋在欧洲。就功能而言，其中只有3栋建筑是办公建筑。因而，未来的最高建筑不仅在分布区域范围上会发生变化，而且建筑高度也会不断突破。预计2010年，建筑高度世界排名前100的建筑叠加在一起的高度会比2006年增高超过5 km。

二、可持续发展趋势

人工环境的营建是影响全球气候变化的主要因素，这点是被公认的。据推算建筑在建设、运营和维护时，大约要消耗其所用能源的

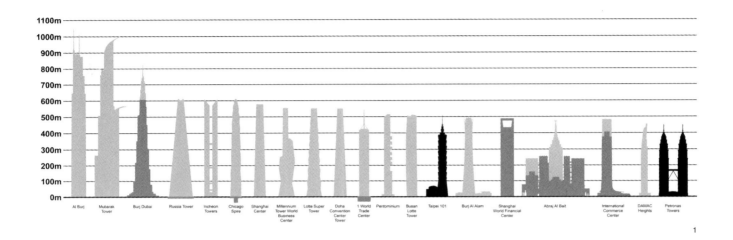

1 2020年20栋世界最高建筑（Copyright CTBUH）

50%，其排放的导致气候变化的气体占全球总气体排放量的50%。在此背景下，国际社会对高层建筑是否具有可持续性、能否成为我们现在和未来城市中正确建筑类型的问题，还没有统一的结论。有人相信，通过集中人口达到的高密度性（由此可减少交通费用，控制城市与城市郊区的扩张），加上建筑高度带来的经济性，使高层建筑这种形式从本质上成为可持续发展的一种设计选择。另外一派则认为，增加建造高度所消耗的能源，加之高层建筑对城市区域产生的影响，使它们从本质上就与环境对立。很多业主、开发商和涉及高层建筑开发的专业人士均陷于这场争论中，至今尚无定论。

目前大多数高层商业塔楼，在国际范围内均遵循一种标准设计模式——直棱、直角，带空调的玻璃"盒子"。这种模式下的高层建筑与场地间没有形成特殊的联系，所以它们可被输送到世界的任何城市。不过也有很少的居住塔楼尝试避开这种只将高效楼层平面竖向堆放在一起的做法。

在过去几十年中，越来越多的专业人士和组织，将适当的环境回报作为高层建筑设计的主要推动力。这种新的设计方向，现正迅速地扩大其影响力，以适应当代对可持续性建筑的急切需求。如位于纽约的美洲银行大厦(Bank of America Tower)计划建成南美洲第一座获得LEED（绿色建筑评估体系）白金认证的高层办公建筑（图2）。大量自然光的引入，雨水收集系统、地送风系统的使用，热电联产、储冰系统的配置及许多其他的科技措施的应用，使这栋办公楼只消耗了同体量建筑50%的能量和用水量。迪拜DIFA Lighthouse Tower的建设目标是达到LEED的白金认证，同时英国伦敦桥中心也以BREEAM（英国建筑研究所环境评估法）"Excellent"标准为目标。

三、低、零二氧化碳的能源制造趋势

近期高层建筑的另一发展趋势是以用地周围低、零二氧化碳的资源制造能源。尽管其中有很多技术仍处于试验阶段，但越来越多的已设计、完成的项目使用了风车、太阳能板、热电联供、冷热电联供、燃料电池和地热泵等系统，降低建筑整体能耗。在能源制造领域，最令人兴奋的是麦纳麦巴林世界贸易中心的建造（图3）。其类似机翼的平面形态使通过建筑间的海风得以加速，并直接吹到直径为29 m的风车上。这样，从整体上可节省11%—15%的电力消耗量。其他建筑，如广州珠三角大厦和伦敦Castle House也使用建筑结合风车的方式，在用地上制造干净的能源。

分析这些趋势产生的原因，我们认为是由于世界范围内城市管理者在法规、规范上的变化。如芝加哥市制定了加速绿色建筑获得许可的程序，意味着为能体现可持续优点的建筑提供更快、更便捷的规划申请步骤。迪拜也制定了绿色建筑规范，要求所有新建居住和商业建筑均须符合国际通认的环境保护准则。与此同时，伦敦规划要求所有的新兴发展项目应该能利用现场可再生能源，创造出相当于10%的城市标准能源。

面对世界范围内高层建筑的建设大潮，特别是可持续性理念的不断发展，有一种倾向认为高层建筑已发展到最先进的时期，其实不然。在可持续设计方法和建造技术开始运用到高层建筑中时，要想使摩天楼"真正"达到可持续的目标，我们仍任重道远。建筑材料所包含的能源消耗和碳释放量，结合在空调、照明和垂直交通方面的高运行能源消耗量，意味着高层建筑必须抓住每个机会去减少能源消耗并生产清洁能源。利用建筑高度，通过风、太阳能和其他方法进行能源生产，其潜力不容忽视。未来高层建筑的最低目标应是"净零能源消耗"，即建筑在用地上利用可持续资源产生的能源与日常消耗量等同。而更高的目标则是真正地中和碳释放量，这就需要创造出能源的盈余量，以平衡从建筑施工、维护到逐渐瓦解，最后只剩废弃结构的整个建筑过程中所包含的能源消耗和碳释放量。

第二个挑战是关于高层建筑设计语汇的，特别是一座高层建筑与其所处城市区域间的关联。从形态上看，很多高层建筑如同高效平面

2 纽约银行大厦（Copyright dbox for Cook + Fox Architects LLP）
3 麦纳麦巴林世界贸易中心（Copyright Robert Lau / CTBUH）
4 芝加哥 Sky Farm（Copyright Won Woo Park / Illinois Institute of Technology and CTBUH）
5 伦敦 Sports Tower（Copyright Patrick Graham & Glyn Lloyd Jones / University of Nottingham and CTBUH）

在垂直空间上的拉伸，或是孤独的城市雕塑，虽然与城市背景的关系只是视觉上的，但高层建筑通常呈现一种专横的姿态，孤独地耸立，不具备任何场所特性。这种构筑模式已敲响"同质性"警报，成为一种"放之四海而皆准"的摩天楼"垃圾"。因而，未来高层建筑与其所在场所间的关系应超越"场所"同义词的内涵，其设计灵感应受到场所物质条件和环境特点的启发。

第三个挑战存在于高层建筑的功能布局中。不论是在建筑内部还是在城市范围，为营建真正具有活力的综合设施，高层建筑必须要创新并超越那些标准的功能设置，包括办公、住宅和酒店，这些标准功能占据了全球高层建筑面积的95%。CTBUH和联合分支机构高层教研组研究了其他的设计方法。不仅得到了富有创新性的建筑形式，而且还创造出新颖的建筑功能。Sky Farm项目能帮助缓和因农业进口（还有接续的粮食运输）导致的环境问题；Sports Tower安排体育功能，其中的游泳池也可被用作谐调液体阻尼器；Solar Thermal立面上的太阳能遮阳板可作攀岩墙；Water Tower内部的竖向储水池能收集雨水且实现循环再利用的最大化效能激发，建筑内部还可以设置风能、太阳能农场（图4~图7）。

在英国和美国大城市管理者已意识到，20世纪曾被广大市民追捧的市郊生活模式，不仅影响城市中心的发展，而且随能源消耗的不断增长影响全球气候环境。内城则因生活密度的相对集中、多样化的空间选择，而被越来越多的人广泛接受。

如果内城所提供的居住空间是保证城市未来发展的关键，在城市核心区地价逐渐增高的情况下，高层建筑已成为可灵活使用的策略以应对城市人口的激增。从表面上看，在英国和美国曾有很多成功的范例。从利兹到利物浦、迈阿密到芝加哥，英国曼彻斯特这座巨大城市的中心已由20世纪90年代的不足100人，在10多年后增长为1.5万人。

深入挖掘这些数据时，你会发现新城市居民中的大多数重新居住在城市中心，他们由两种社会经济群体构成——年轻的单身人士或新婚群体，还有就是"空巢"一族及年纪较大的退休人群，他们对面积较大的市郊住宅没有需求，反而想更方便地利用城市中的各项设施，如餐馆、剧院和公园。其他的社会经济群体属真正的、长期的往返于市郊与城市中心（大多数为家庭）的人口，他们依然住在市郊并继续着从城市向市郊的迁徙。

目前建成的高层建筑在不断收缩的城市人口面前，变得无足轻重。事实上英国在第二次世界大战后迅速建成的高层住宅，大多是不受欢迎的，因为其建筑形式并不适合家庭。流行的小型公寓单元主要服务于单身和无子女的夫妇，当然，这一部分人口也寻求这种公寓。不过，也不一定要陷入这种局限中。高层建筑仍有重新发掘自身的机会，它可为密集的可持续性城市及在城市中生活的人提供理想的解决方案。高层建筑被视为不适合居住的主要原因是缺少开放、休闲、交往的空间，如街道、步行道、广场和公园等。我们可以在建筑纵向空间创造如空中公园和广场这样的空间，既增加安全性又营造舒适性。

从开发者的角度而言，在高层建筑中营造没有收入回报的开放空间，花费实在太高。不过高质量的设计品质及近期的可持续设计理念，已成为建筑创收的突出因素。人们越来越能接受为高质量设计和改善的环境，特别是带可持续性证书的建筑，付额外费用。举个例子，2006年由McGraw Hill Construction在美国对开发商进行的调查显示，可持续建筑的出租率要比一般高层建筑高出3.5%，房租水平也相应提高3%。

在大量建成的居住案例中，竖向的空中花园和交流空间在国际范围均被认可。如新加坡的Duxton Plain Housing、芝加哥340 On the Park和澳洲黄金海岸的Q1塔。但是一项来自商业办公世界的案例能更清楚地显示未来趋势。比起通常大家所期待的、楼层平面效率应在70%以上的指标，法兰克福德国商业银行的楼层平面效率只有50%，仿佛是开发商的"噩梦"（图8）。不过在失去的办公面积中形成的与建筑齐高的中庭、围绕建筑的半开敞空中花园及与工作台相距不过

6　　　　　　　　　　　　　　　　7　　　　　　　　　　　　　　　8

6m、7m的能开启的窗，均为此建筑营造出高质量的绿色交互空间。建筑要进入商业市场，其内部环境的质量越高，每平方尺所带来的收益就越大，同时还能通过改善环境提高工作效率。

要真正实现具有社会意义的可持续城市议题，未来高层建筑的发展应在环境、设计和功能方面做出更好的回应。作为一种建筑形态，为实现可持续的目标，高层建筑需重新发掘自身所蕴含的意义——高度集中的生活、工作、娱乐中心，带有创新的形式、技术和环境，去面对未来气候变化的挑战。

参考文献

[1] ASTON A. Bank of America's Bold Statement in Green[J]. Business Week, March 19, 2007.

[2] DALEY R M, JOHNSTON S. Chicago: Building a Green City. Proceedings of the CTBUH 8th World Congress "Tall & Green: Typology for a Sustainable Urban Future" [C]. Dubai: March 3-5, 2008: 23-25.

[3] FOX R F. Provocations: Sustainable Architecture Today[C]. Proceedings of the CTBUH 8th World Congress "Tall & Green: Typology for a Sustainable Urban Future". Dubai: March 3-5, 2008: 354-361.

[4] IPCC. Climate Change 2007: The Physical Science Basis. Summary for Policy Makers[M]. Cambridge, United Kingdom, New York, USA: Cambridge University Press.

[5] OLDFIELD P. The Tallest 10 Completed in 2007[J]. CTBUH Journal, Issue 1, 2008: 16-17.

[6] OLDFIELD P. The Tallest 20 in 2020[J]. CTBUH Journal, Fall 2007: 24-25.

[7] SMITH A. Burj Dubai: Designing the World's Tallest[C]. Proceedings of the CTBUH 8th World Congress "Tall & Green: Typology for a Sustainable Urban Future". Dubai: March 3-5, 2008: 35-42.

[8] SMITH P. Architecture in a Climate of Change: a guide to sustainable design[M]. Architectural Press: Oxford 2nd Edition.

[9] SMITH R F, KILLA S. Bahrain World Trade Center(BWTC): The First Large Scale Integration of Wind Turbines in a Building[J]. The Structural Design of Tall and Special Buildings, No.16, CTBUH 1st Annual Special Edition, 2007, John Wiley & Sons: 429-439.

[10] WEISMANTLE P A, SMITH G L, SHERIFF M. Burj Dubai: An Architectural Technical Design Case Study[J]. The Structural Design of Tall and Special Buildings, No.16, CTBUH 1st Annual Special Edition, 2007, John Wiley & Sons: 335-360.

[11] WOOD A. Green or Grey? The Aesthetics of Tall Building Sustainability[C]. Proceedings of the CTBUH 8th World Congress "Tall & Green: Typology zzfor a Sustainable Urban Future. "Dubai: March 3-5, 2008: 194-202.

6　芝加哥 Water Tower（Copyright Steven Henry & Hannah Cho / Illinois Institute of Technology and CTBUH）
7　芝加哥 Solar Thermal Tower（Copyright Thomas Denney & Bradley Weston / Illinois Institute of Technology and CTBUH）
8　法兰克福德国商业银行（Copyright Foster+Partners）

作者简介

安托尼·伍德　美国芝加哥高层建筑与城市环境协会，芝加哥伊利诺伊伊理工学院建筑系教授

菲利普·欧德菲尔德　英国诺丁汉大学建筑环境学院副教授

李靳　凯盛国际（上海）有限公司设计师

汤岳　英国诺丁汉大学建筑环境学院博士

THE RESEARCH OF INTERACTION BETWEEN SKYSCRAPERS AND URBAN SPACE
超高层建筑与城市空间互动关系研究

梅洪元　梁静 | Mei Hongyuan　Liang Jing

从第一幢超高层建筑拔地而起的那一刻开始，有关它的争论就从来没有停止过。美国9·11事件之后，超高层建筑又一次成为人们讨论的焦点，整个世界对超高层建筑心生恐慌，其存在价值遭到质疑，反对和限制建设超高层建筑的呼声一浪高过一浪，理由更是五花八门、莫衷一是：形象工程、资源浪费、破坏城市景观、威胁城市环境、引发灾难、影响城市交通、存在巨大的消防安全隐患……

然而，超高层建筑并没有在人们激烈的争论声中停止前进的步伐，不仅数量没有减少的迹象，而且高度不断突破极限：2004年中国台北国际金融中心落成，使超高层建筑的高度一举突破500 m大关，而高达610 m的广州新电视塔和日本东京新电视塔以及阿联酋800 m的迪拜Burj Dubai大厦相继建设，这些建筑的落成又使超高层建筑成为世人关注的焦点。

通过分析超高层建筑引发的城市问题（表1），我们可以看到，问题产生的主要原因是超高层建筑与城市空间没有实现有机结合。因此，本文借鉴结构主义的观点[1]，重点研究超高层建筑与城市空间的关系，探讨二者之间的相互作用机理，揭示其中的规律，并试图建构一个相对普适的关系评价框架，为两者之间建立积极的互动关系提供一种导引。

一、超高层建筑与城市空间的互动关系

超高层建筑的建设、使用以及自身的形体会作用于城市空间，尤其体现在超高层建筑的城市集聚作用及其对城市空间视觉方面的影响。超高层建筑以城市为背景，与城市空间的结构、景观、功能以及环境产生互动性关联，这些关联要素影响着超高层建筑的发展，同时超高层建筑的发展也反映并促进了这些要素的发展。

1. 与城市空间结构的互动

首先，超高层建筑的布局与城市空间格局的变化紧密相连。城市不断发展进化，因此其空间结构也是动态变化的。埃里克森（Rodney A. Eriekson）将城市空间结构的发展过程归纳为集中——分散——再集中的周期运动，此过程以"集中"为特征，其中的分散过程可以理解为在另一个区域的集中，超高层建筑的建设在这一过程中也可以被看作一种集聚的方式，即"为了分摊高昂地价的垂直性集聚"[2]。随着城市土地的日益稀缺，这种以超高层建筑建设为主的垂直性集聚已经成为形成城市中心与副中心的主要方式。

其次，城市中心区超高层建筑的发展，可以促进城市功能的自我调整，发挥城市空间结构的自组织机制优势，有效促使城市空间结构在新层次上的发展和自我完善，例如超高层建筑使费城CBD沿商业大街向西平稳移动，使纽约CBD核心区自百老汇向北移至23街，第五大道成为移动轴。同时，超高层建筑又对CBD具有阻滞和逆转作用，例如克利夫市1927年竣工的213 m高的顶点大厦，使几十年一直向东移动的CBD发生逆转而在公共广场附近保持40年之久；20世纪30年代建成的洛克菲勒中心，成为纽约城CBD旧核心之外的副核心。

2. 与城市空间景观的互动

首先，超高层建筑与城市空间景观的互动表现在城市天际轮廓线上。由于天际线是城市整体面貌的垂直空间投影，反映的是城市建筑的总体轮廓，因而超高层建筑的高度及密度分布成为控制天际线的两大主要指标。建筑高度与天际线有必然的联系，但并不是高度越高，天际线特征越明显。对天际线有明显影响的是建筑体之间的高差（图1），同时，建筑密度也在一定范围内影响天际线的趋势和走向，超高层建筑的密度过大会影响天际线的轮廓，并掩盖建筑单体的特色。比如上海浦东陆家嘴地区，由东方明珠、国际会议中心、金贸大厦等建筑构成的浦江东岸天际线目前十分清晰，但随着未来该地区筹建中的超高层建筑相继落成，其轮廓线的艺术美感必将因过高的建筑密度而大大降低（图2）。

1 上海陆家嘴天际线（图片来源：www.skyscrapercity.com）
2 上海外滩天际线，2008年（图片来源：www.skyscrapercity.com）

表1 超高层建筑引发的城市问题列举

城市问题	表现特征
布局矛盾	超高层建筑密集区建筑密度过高、绿地较少、日照的矛盾突出，非密集区则随机分布、没有秩序，有损历史风貌、破坏自然地景形态
屏风效应	高楼密集成片犹如天幕，会危害局部物理环境和景观环境，影响周围居民的生活环境、空气质量、景观及自然采光，令居民感染呼吸道系统疾病的比例增加
过度标志	超高层建筑争当城市空间中的主角，导致标志性建筑泛滥，城市空间秩序混乱、缺乏整体感，相互之间缺乏必要的功能联系，相邻建筑的公共设施缺乏统筹规划
特色危机	城市历史文脉失落，许多有价值的风貌荡然无存，缺乏特色、面貌雷同的现代化摩天大楼以及大都市的巨大尺度中使人们不知自己置身何处，丧失了归属感和家园感

其次，超高层建筑与城市空间景观的互动还表现在景观视廊的创造与保护上。景观视廊是城市景观资源的重要体现，形成通畅的景观视廊是城市设计的重要原则之一。不同城市格局中景观视廊的表现形式也有所差异，有的是自然环境的融合，有的是历史节点的汇聚，但无论何种形式，超高层建筑都会与景观视廊的强化与整合产生强烈互动。处在景观视廊序列中的超高层建筑必须自觉服从和服务于城市空间塑造的整体要求，其造型设计和空间布局应有利于序列感的形成，例如巴黎德方斯拱门前的超高层建筑造型都较为简洁，而且大都布置在轴线的两侧，既是城市轴线的尾声，同时也衬托了德方斯拱门的中心地位。

最后，由于规模大、体量显著，超高层建筑的总体布局对区域城市空间会产生较大的影响。因此超高层建筑的总体设计，须首先积极研究周边的建筑形态及该区域城市空间的城市设计导则，选择合理的布局方式，与周边的建筑共同构成连续、积极的区域城市空间。

3. 与城市空间功能的互动

超高层建筑与城市空间功能的互动表现在对于城市功能的优化上。根据西方城市学家将城市功能分为基本功能（即外部功能）和非基本功能（即内部功能）的理论基础，我们将这种优化分为内部功能的优化和外部功能的优化。城市外部功能是指一个城市对本市以外区域提供服务的功能，即其作为区域中心的作用，是城市形成和发展的原动力。城市的内部功能是指城市为其内部的企事业单位、社团和居民服务的效能，表现为城市自身的凝聚作用，是维持城市自身正常运转的保证。

内部功能的优化包括城市运行效率的提高以及居民生活方式的丰富。超高层建筑依靠内部多种功能的综合化、功能组织模式的多样化以及建筑交通的一体化大大提高了城市内部系统的运行效率，同时也改变了人们的生活环境和生活方式。超高层建筑如果要实现丰富居民生活的目标，必须积极开发和创造多种空间组织模式，同时维持街道的氛围和形态，从而使城市保持活力。

外部功能的优化包括城市经济的促进、城市竞争力与政治影响力的提高等。超高层建筑是时代的产物，不能脱离政治、经济和技术的发展。首先，它对于城市经济的发展具有促进作用，比如上海超高层建筑的建设十分显著地带动了经济的发展，楼宇经济已成为上海市中心各区的主要税收来源；其次，许多成功的超高层建筑策划能够提升城市的政治影响力和城市形象，从而实现"一幢楼带起一个城市"的触媒效应，比如马来西亚的佩重纳斯双塔标榜了民主政治的成就和民族自信心，CCTV中央电视台新址大厦的建设则彰显了中国改革开放所取得的辉煌成就。

4. 与城市空间环境的互动

超高层建筑与城市空间环境的互动包括两个方面——与城市人文环境的互动和与城市物理环境的互动。一个城市的人文环境由许多因素综合形成，超高层建筑作为物质载体能够反映出城市的文化内涵和社会文化心理，其实现途径包括外部形象的文化表达、空间组织的内

表2 超高层建筑与城市空间互动关系评价框架

一级指标	二级指标	主要评价内容	评分				
			好（2）	较好（1）	不足（0）	较差（-1）	差（-2）
城市空间结构	集聚与扩散	城市集聚作用力					
		城市扩散引导力（城市发展方向的引导力）					
城市空间景观	宏观尺度	天际线的秩序感					
		城市景观视廊的创造与保护					
	中观尺度	重要景观节点的形成与保护					
		区域城市空间景观的维护					
城市空间功能	内部功能影响	运行效率提高度					
		生活方式丰富度					
		群众满意度					
	外部功能影响	楼宇经济效益率					
		提高城市竞争力					
		政治影响力					
城市空间环境	人文环境	文化体现度（城市特色）					
		群众认可度					
	物理环境	基础设施改善度					
		自然环境影响度（越小越好）					
总分		以上各项分数相加					

注：评分标准（好——优势所在；较好——有待改进；不足——亟需改进；较差——需花很大力气才能补救；差——难以挽回）

涵彰显、技术与情感的动态平衡以及城市文脉的开拓创新。借助超高层建筑的庞大体量，城市的文化内涵被成倍放大，并传递到城市各个角落，而那些经久不衰的建筑也必然是在文化表达上最大限度地获得群众认可的建筑。

城市物理环境是城市文化环境的基础保障，高尚的人文环境离不开舒适的物理环境。超高层建筑的建设通常会使城市的基础设施水平、局部区域的城市交通网络和市政工程质量得到改善，但却很难避免对城市自然环境的负面影响，因此在普遍追求绿色建筑的今天，超高层建筑的生态化设计受到前所未有的重视，减少其对自然环境的影响是每个建筑师都应肩负的责任。

二、超高层建筑与城市空间互动关系评价框架

1. 价值与理念

评价超高层建筑与城市空间互动关系的强弱，除了关注其视觉影响，更重要的是探究两者在功能、经济方面的互动机制，需要我们从城市空间的结构、景观、功能、环境等各方面进行全方位考虑。一直以来，建筑领域对两者关系的分析缺少综合的评价方法，而一个相对普适的超高层建筑与城市空间关系评价框架可以较好地发挥作用。

2. 互动关系评价

基于以上的分析，我们尝试建立一个评价框架，以超高层建筑与城市的互动关系为评价对象，其关系的强弱为评价结果，评价内容涉及经济、社会、政治、功能、环境、管理经营等多个方面，即有定量的评价。评价的总分被划分为三个层级——负分、零分、正分，分别表示关系弱、中、强（表2）。

三、结论

超高层建筑与城市空间关系的研究是一个理论与实践相结合的课题，涉及的内容也极其庞杂，从不同的角度出发，我们会看到超高层建筑与城市空间存在着不同的关系类型。本文尝试在理论层面为其建构一个探索性的研究框架，以期引起有关方面和同行的更多关注。在此基础上，从横向、纵向两个纬度进行更深化的课题研究。

注释

① 特伦斯·霍克斯在《结构主义和符号学》中指出，"结构主义认为，事物的真正本质不在于事物本身，而在于各种事物之间的，并为人的认识所感觉到的那种关系"。世界是由各种关系而不是由事物构成的观念，是结构主义者思维方式的第一原则。根据结构主义观点，决定形式的主要因素是形式内部的组织关系，而不是取决其构成元素。

② 沙里宁在谈到城市的空间集聚时，精辟地总结了城市四种基本的集中方式，即为防御战争的强迫性集聚、为追求经济效益的投机性集聚、为分摊高昂地价的垂直性集聚以及为信仰目标及交往活动需要的文化性集聚。

参考文献

[1] 朱喜钢. 城市空间集中与分散论 [M]. 北京：中国建筑工业出版社，2002.
[2] 孙志刚. 城市功能论 [M]. 北京：经济管理出版社，1998.

作者简介

梅洪元　全国工程勘察设计大师
　　　　哈尔滨工业大学建筑学院院长、教授、博士生导师
　　　　哈尔滨工业大学建筑设计研究院院长、总建筑师
　　　　《城市建筑》主编

梁静　　哈尔滨工业大学建筑学院讲师、博士

SKYSCRAPER IN THE VIEW OF URBAN DESIGN
城市设计视野中的高层建筑

卡瑟利娜·鲍尔斯　黛娜·哈拉萨　| Katharina Borsi　Dana Halasa
付斌　译
汤岳　校审

在世界范围内，城市正经历高层建筑发展的快速膨胀期，伴随城市密度的增加和竖向高度的攀升，新的天际线塑造出新的城市景观。这种因高层建筑而促动的城市发展源于多种因素的推动力：城市生长、地价攀升、商业机遇、企业要求、土地稀缺、社会需要甚至是对都市标志性的需求。与之相应，很多城市也在探索将这种竖向延展的空间融入现存城市肌理中的正确方法。

对高层建筑而言，缜密周到的设计方法也需顺应城市的肌理和个性，还应考虑高层建筑与城市设计交叉部分存在的普遍问题。本文结合类型学的有关理论，通过对伦敦城市中几座高层建筑案例的研究，分析它们与城市环境和场地间的关系，所涉及的问题均属城市设计的重点，同时也是高层建筑类型学该强调的内容，其中包括高层建筑对城市天际线的突出作用、高层建筑首层与城市公共空间及更大区域范围城市空间的相互作用（图1）。

一、《伦敦计划》的空间策略和指导原则

鉴于当地和全球范围内的经济增长，过去的20年间伦敦发生了"戏剧性"变化。据预测，在下一个20年，这座城市不得不多容纳50万人口，随之而来的是对办公空间的大量需求。据"伦敦计划（2004年后的改造计划修订稿）"所制定的战略性空间规划，城市发展应以维护和提升伦敦城市环境为目标，在不扩张用地面积、不占用现存绿化带、不进一步蚕食伦敦内部绿色空间的前提下，合理地消化增加的人口。此外，市政府强调实现这种和谐增长要以提升市民生活质量和维护城市特色为目标。

发展中的高层建筑被视为伦敦空间战略的重要组成部分，这就意味着伦敦希望更紧凑而非横向扩张的城市。相应地，市政府也力图促进高层建筑的建设，使其成为具有魅力的地标建筑，强化城市个性特征，同时也为因相关活动而聚集的经济组团提供极富凝聚力的场所。

伦敦计划建议所有的大尺度建筑，包括高层建筑，应依照最高标准设计，使其不论从哪个角度观看，均是具有魅力的城市元素，进而构筑极富韵律的城市天际线，为人们提供视觉上的焦点。此外，高层建筑只有在不阻碍视觉走廊的前提下才能获规划许可，这些视觉走廊是为了确保纪念性构筑物和地标性建筑的中、长观赏视线而建设的，其中最著名的是圣保罗大教堂视觉走廊。

当然，高层建筑也为伦敦增添很多绝佳的观景机会。许多高层建筑的顶层设有受人欢迎的公共空间。最大限度地开发高层建筑在用地方面的潜能，这一目标推动了高层建筑高质量的设计及公共场所氛围的营造，也要求高层建筑的尺度比例、空间构成与其他建筑、街道、公共和私密空间及各种城市景观元素，乃至周边环境背景相协调。

伦敦城市的独特之处在于许多不同类型、不同年代的建筑和空间相互毗邻。在这里，建筑和场所不应是孤立的，周围环境背景通常对建筑个性特点的营造是很重要的，设计师应树立正确的观念对建筑遗产予以保护。我们也注意到，伦敦计划对高层建筑的设计制定了一系列空间和形式上的政策和建议。这些政策和建议只是泛泛的定义，为建筑师的诠释留下广阔空间。我们不应仅注重视觉方面的效果，而忽视高层建筑作为更大尺度的城市元素，对街道和区域等空间氛围的营造所带来的贡献。

Development Securities PLC编写的报告"高层建筑：未来的构想还是过去的受害者"，已开始探讨这个问题。在作者眼中，"高层建筑如何与地面交接"是和"如何与天空相交接"一样重要的问题，如果高层建筑要在伦敦的城市发展中扮演更为重要的角色，它就必须从头至尾地整合于城市的肌理中。如果这一点能够实现，更加高效和富有视觉连贯性的伦敦也会在现有城市范围内继续成长、不断更新，增强其在全球范围内的竞争力。

1 托特汉姆法院路上的高层建筑（摄影：Dana Halasa）
2 托特汉姆法院路上的中心塔（摄影：Dana Halasa）
3 高层建筑底层与城市空间（摄影：Dana Halasa）
4 汇丰银行大厦（摄影：Dana Halasa）

这份报告还为高层建筑在城市中的实施提出了更积极的规划策略。报告指出，目前高层建筑过分强调建筑高度的作用及对天际线的作用，而忽略对街道层面的考虑，其残留的开放空间对公共领域也没有贡献。当"最高设计质量"被视为建筑必须满足的标准时，我们只能通过将建筑完全融入城市肌理才能全面实现这一目标。

报告强调应仔细考虑高层建筑边缘与街道的关系，指出从人行道边缘到活动频繁的建筑正立面，应保持最低限度的建筑退线，利用不断出现的门窗等构件增强建筑与公共领域间的互动。此外，内与外、公共与私密领域的区分在这里应被模糊处理，将城市公共空间结合高层建筑共享大厅或顶层空间设置。通过交通流线、室内公共空间和开敞的入口，将街道活力与建筑内在活力紧密连接，使高层建筑或高层建筑群，从实体结构、感官效果等方面均能与周围环境相结合，增强现存街道和步行空间的联网（图2）。这些观点并不是让高层建筑逃避地标的作用或构筑城市天际线之类的重要功能，而是提供更多、更具有针对性的类型学建议。通过这些建议，高层建筑可在更大城市区域内做出贡献，而不仅局限于自身。在这里，高层建筑设计对不同尺度的环境均有影响，不仅包括与建筑紧密相连的公共空间，还包括街道和街区层面。在下面的案例研究中，我们将探讨高层建筑在城市形成过程中的积极作用。

二、托特汉姆法院路上的高层建筑

托特汉姆法院路是城市中心的交通动脉，北临优斯顿路，与伦敦最繁忙的购物街牛津街相交，临街有售卖电气产品和高端家具的商铺，人行道和交叉口均有大量人流。

1. 中心塔 Centrepoint

在托特汉姆法院路、Charing Cross路与牛津街交叉口耸立着"Centrepoint"——一栋建于1963年的地标大厦，建筑高117 m，可容纳35层的办公空间，独特的混凝土肌理使其成为伦敦地标建筑（图3）。但由于首层平面设计的不妥，建筑不仅未对周围剩余空间做出贡献，且还塑造了不友好的环境。该建筑在街道上并没设入口，而是通过升起的平台引导人流进入，这就为行人的运动和进入制造了障碍。由于Centrepoint北面和西面街道的界限模糊，公共领域未被围合，该建筑也未能延续周边建筑物的横向组织模式。因而，在建筑首层缺乏互动的空间形式和功能设置，导致建筑物剩余的开放空间既简陋也不友好。

2. 汇丰银行大厦

相比之下，沿托特纳姆法院道再向北的汇丰银行大厦，则是高层建筑与公共领域和谐对话的成功案例。该建筑采用现代混凝土结构，重新翻修的外立面呈现为由石材和玻璃构成的内敛沉静的气质。建筑裙房从相邻建筑构成的连续边界线上后退，最终形成虽小但充满活力的"袋型"公共空间。高层塔楼建筑则进一步后退，以减少对街道的压迫感。三层的裙房缓解了该建筑与周围环境在尺度上的差异，与建筑外围连续布置的店铺相结合，形成强烈的互动性（图4）。

3. 优斯顿大厦

托特汉姆法院路北端的优斯顿大厦是人们的视觉焦点。虽然优斯顿路的8条车行线将大厦与托特汉姆法院路割裂开，但其灵活的地下交通、突出的高度及优越的地理位置，仍确保其成为独具魅力的地标性建筑。

大楼的成功应归功于对建筑底层的重新开发，这也是Triton广场规划的一部分。Triton广场由建筑师Sheppard Robson进行总体规划，通过一组建筑围绕新的广场构成城市的公共空间。这项开发首先要实现建筑形式上的统一，还要确保建筑物围绕一系列连续的开放空间进行设计。虽然这些构想只是部分地实现了，但是优斯顿大厦一层的连续铺面和光鲜建筑外观，仍旧还是能吸引人流进入并且穿过广场

5
6

到达毗邻的建筑（图5）。

以上这三栋高层建筑在整合周边开放空间方面各有千秋，从宏观尺度来看，这三栋地标性建筑帮助定义和强化了托特汉姆法院路的空间形态。它们并没有在街道尺度上喧宾夺主，而是界定区域范围，进而增进视觉呼应。Centerpoint和优斯顿大厦构成了托特汉姆法院路两个端头的视觉标注，而汇丰银行大厦则帮助三栋建筑间长距离对话的展开。托特汉姆法院路同时连接西侧Fitzrovia区和Bloomsbury区，Centrepoint南面是科芬花园，街对面是Soho住宅区。考虑到每个区域都有自己的建筑形式、空间布局和功能特点，位于同一条街上的三栋高层建筑间就产生了内在联系和视觉协调，强化了托特汉姆法院路的城市主干道功能。

三、伦敦城的高层组团

伦敦城位于泰晤士河北部，现已成为继卡纳瑞码头之后高楼大厦主要的聚集地。我们研究的重点是三个成组团布置的高层建筑——理查德·罗杰斯的劳伊德大厦（1986）、福斯特和合伙人设计的瑞士再保险公司总部（2004）及理查德·罗杰斯事务所设计的Leadenhall街122号（预计2011年完成）。虽然将规划阶段的Leadenhall街122号拿到文中做统筹考虑，使分析变得不确定，但在城市设计层面还是颇具意义的。

1. 劳伊德大厦

劳伊德大厦是伦敦最受欢迎的竖向地标，其充满韵味、情感丰富的立面效果，闪亮的金属表皮，面向圣海伦广场的外部电梯及南面中庭光滑的穹顶，向人们传达一种机械美学，其构造精致的立面表达及元素的合理分配，也减少了视觉上的紧张感。

劳伊德大厦对北侧圣海伦广场的界定作用主要体现在视觉方面。建筑入口设在街道上，零售店铺和街道层面上的建筑界限基本都是封闭的。但是，位于建筑边缘的一系列踏步可通向Leadenhall市场——一个具有历史意义且极受欢迎的建于19世纪的室内经营场所。该市场的鹅卵石步行道和玻璃屋顶，使其成为充满魅力的购物、餐饮空间，每当午餐时间和下班后，这里便成为非正式会议和社交的重要场所之一（图6，图7）。在劳伊德大厦的背面，建筑配合一系列的高差变化营造活跃的街道空间，创造出伦敦城内最具活力的聚会场所。

2. 瑞士再保险公司总部

40层的瑞士再保险公司总部大楼凭借弯曲的几何外形成为伦敦最负盛名的地标性建筑。该建筑形式以一定半径产生的圆形平面为基础，由基础逐渐向上放大，再依次向顶点缩小。建筑在靠近底部时变纤细的处理方式，是针对城市传统街道空间狭窄的创新性解决方法。其适中的平面曲率意味着减少建筑体量的压迫感。置身街道中，行人几乎忽略了大厦的存在，直到他们到达广场边缘，才注意到其高耸的体量。广场的活力则通过建筑底部排列的商业和休闲单元营造出来。但在设计中，这些单元被放到建筑表皮后面，弱化了建筑内部和广场间的视线联系。因而对比那些用地边缘种植的成年大树和低矮石墙点缀形成的休憩空间，该广场缺乏凝聚力，没有劳伊德大厦背面的袋状空间的亲密感，也没有圣海伦广场的活力。但相对于更广泛的城市背景环境而言，大厦的贡献则在于另一个公共聚点的形成——由Leadenhall市场的公共空间和广场开始，到Leadenhall街122号，贯穿圣海伦广场最后进入瑞士再保险广场（图8）。

3. Leadenhall街122号大厦

Leadenhall街122号大厦高222 m，凭借其独特的几何形状必将成为伦敦天际线中的突出形象。其建筑高度、圆锥形外皮是通过研究场地与现存建筑间的关系确定的。

Leadenhall街122号可被视为在瑞士保险公司总部大厦基础上的深入发展。正如建筑师所说："建筑底层沿对角线层层退后，创造出很大的公共空间且向南面开敞。这种尺度宏大的、半封闭式的、类似教堂内部的建筑空间在伦敦还没有先例，这里将会形成重要的会议场所和独一无二的休闲胜地。巨大且微倾的建筑体量将圣海伦广场与Leadenhall街连接起来，使空间流畅连通。大厦不仅通过首层空间与

5 优斯顿大厦（摄影：Dana Halasa）
6 劳伊德大厦底层空间（摄影：Dana Halasa）

7

8

街道和相邻广场形成一种主动的内在关联，更明确围合出一个7层楼高的公共空间，表达该建筑与圣海伦广场、Leadenhall街在空间和功能上的连续性。这幢建筑就这样既保持着原有的步行连接体系，还在伦敦城内营造出新的会议场所。

我们对劳伊德大厦、瑞士再保险公司总部大厦和Leadenhall街122号大厦首层平面的描述，是为说明高层建筑未来的发展趋势，即从建筑形式、空间及功能安排上与周围环境形成整体，这种演变趋势也应归功于市政府在伦敦计划中导则和政策的成功制定。

这些建筑组团彰显了伦敦城的个性和特点，以相对较小的占地面积，延续城市现有的中世纪的布局模式，使人流穿越其中。伦敦城作为承载市场经济发展的场所，其兴起的基础是关系网络和资源交换。这种非正式的经济格局依赖或大或小的凹形公共空间所形成的网络来支持、推动，而这种网络正是以高层建筑的高使用率为基础建立的。

四、结论

综上所述，我们并不是要面面俱到地阐释高层建筑在建筑设计和城市规划方面的种种问题，我们论述重点是探讨高层建筑与城市语境的整合模式。案例的选择也不是为单纯对比高层在功能布局和环境营造方面的差别，而是力图表明其对于城市结构更广阔、更具战略意义的作用。

阿尔多·罗西曾经说过，城市应被理解为是由零部件组成的，每个构件均有自身变化和转化的过程。托特汉姆法院路案例探讨了高层建筑在整合一条道路时所扮演的关键角色——帮助城市空间连接和分割区域的任务。伦敦组团案例则说明高层建筑在空间、经济层面为伦敦城做出的贡献；伦敦金融区的成功部分归功于同步协调、交换模式和准确交流，这些均是通过高层建筑在城市肌理中相对均匀的分配强度来实现的。城市中高层建筑与开放空间的组合，特别是当它们紧密结合起来时，就构成了文化、信息交换和休闲娱乐相结合的基础。

7 劳伊德大厦底层空间（摄影：Dana Halasa）
8 瑞士再保险公司总部大厦（摄影：Dana Halasa）

参考文献

[1] BARTH L. Public Lecture. Architectural Association School of Architecture. London: 12, 10, 2007.

[2] English Heritage, CABE. Guidance on Tall Buildings, 01.08.2008, available at http://www.cabe.org.uk/AssetLibrary/2139.pdf.

[3] CABE Design Review 122 Leadenhall Street, 01.08.2008, available at http://www.cabe.org.uk/default.aspx?contentitemid=845&field=filter&term=london&type=7.

[4] CARMONA M, FREEMAN J. The Groundscraper: Exploring the Contemporary Reinterpretation[J].Journal of Urban Design, 10: 3: 309-330.

[5] Greater London Authority The London Plan. Spatial Development Strategy for Greater London. Consolidated with Alterations since 2004. 01.08.2008,available at http://www.london.gov.uk/mayor/strategies/sds/index.jsp.

[6] LSE Tall Buildings: Vision of the Future or Victims of the Past? A report by the London School of Economics for Development Securities PLC.

[7] Rogers Stirk Harbour + Partners, The Leadenhall Building, 01.08.2008,available at http://www.richardrogers.co.uk/render.aspx?siteID=1&navIDs=1,4,25,361,366.

[8] Rossi A. The Architecture of the City (Cambridge, MA: MIT Press).

作者简介

卡瑟利娜·鲍尔斯　英国诺丁汉大学建筑环境学院讲师，硕士生导师

黛娜·哈拉萨　英国诺丁汉大学建筑环境学院研究生

ANALYSIS OF THE FORM AND PERFORMANCE OF HIGH-RISE BUILDINGS IN THE NEW CENTURY
新世纪高层建筑形式表现特征解析

梅洪元　李少琨 | Mei Hongyuan　Li Shaokun

高层建筑的发展与社会进步和时代变迁密不可分，尤其是进入新世纪以来，一方面随着设计手段与相关技术的不断成熟，高层建筑深入到住宅、办公、商业、学校、医疗等多种建筑类型，同时又形成了多种高层综合体。这些建筑类型和使用功能的多样化、复合化必然会反映在高层建筑的形式上，使其呈现多元综合的形式表现特点。另一方面，在当代高新技术的支持下高层建筑的创作观念也呈现出多元化的趋势，这些观念有的运用理性和夸张的手法，强调标新立异；有的强化新技术的应用，表现技术和材料美感；有的发挥地域特色，融合地域文脉；有的注重城市文化，关注建筑的文化象征；等等。加之各个地区的社会、经济、技术背景各不相同，必然使高层建筑的形式表现呈现出多元化、综合化、高技术化的发展趋向（图1～图3）。

一、形式表现的基本特点

"建筑形式问题是建筑学的基本问题之一，建筑的功能要通过建筑'形式'去实现，建筑的思想、观念、意义也要透过'形式'来表达。事实上建筑形式关联'功能'和'意义'，处于中心地位，而在建筑创作的认识上却又处于从属位置，建筑形式既是实现功能的'工具'，又是表达意义的'媒介'。"因而，建筑作为一种具有使用功能的实用艺术，既有"物质"属性又有"精神"属性，是双重属性与价值的统一。

高层建筑作为一种建筑类型，一方面继承了建筑"物质"与"精神"的双重属性，注重内部功能与意义内涵的双重关联；另一方面它与结构选型密切相关，这在很大程度上影响甚至决定形式，所以功能空间的呈现、环境意义的展现、结构逻辑的显现均成为高层建筑形式表现的基本特点。

1. 功能空间的呈现

功能空间的呈现，即高层建筑外部形式对内部使用功能的一种理性表现。按使用性质划分，高层建筑涵盖办公、旅馆、商业、住宅等类型。在建筑内部，不同的功能空间均有不同的空间组织模式和容量配比。若将内部的功能空间理性地反映到建筑的外部，就会产生各异的建筑形式。

进入新世纪，随城市化进程的不断加快以及社会多样化需求的不断增加，迫切需要在一栋高层内集合多种使用功能，以此提高建筑空间的使用效率和经济价值。因此，高层内部的功能组织往往以多种功能空间的复合为原则，这种复合不是简单的水平分区和垂直分层，而是对功能空间的优化组合，使其形成既相对独立又相互联系、既相互依存又相互支撑的关联方式，创造有机、复合的整体。高层建筑内部多种空间复合的组织模式为外部形式的塑造奠定基础，也为形式表现的丰富性、多样性带来契机。通过将内部空间逻辑地呈现到外部，进而真实地反映各组成部分间的差异性，再通过对比、协调等手法将高层建筑塑造为多样统一的有机整体。

2. 环境意义的展现

环境意义的展现，即高层建筑外部形式对场所环境的积极回应。这里的环境包括城市的各种建筑实体、街区空间等显性要素，也包括城市的时代精神、文化传统、历史文脉、社会风俗等隐性要素。高层建筑作为城市发展的产物，与环境密不可分，自身发展也要依托于环境；同时因其巨大的空间体量、复杂的组织规模以及高效的经济价值，反过来又会对城市产生深刻的影响。在当代全球一体化的趋势下，人们开始更多地关注地域性的发展主题——尊重不同地域环境间的差异，强调不同场所环境的特色。高层建筑因其引人注目的体量与造型而构成区域的标志，它的形式表现往往会成为人们感知城市环境风貌、体验城市特色的重要途径。如位于北京市西城区西长安街的国家电力调度中心，特殊的地理位置对建筑的形式语言提出很高的要求。该建筑也正是通过对场所的积极回应，将中国传统文化元素与现代建筑语言进行有机融合，塑造出既具时代特征又具地域传统和文化精神的建筑品质，同时也能充分提升环境价值，为长安街增添了新的气象与特色。

3. 结构逻辑的显现

结构逻辑的显现，即高层建筑外部形式对内部结构特征的一种逻辑反映。高层建筑的产生与发展离不开结构技术的支撑，恰当的结构

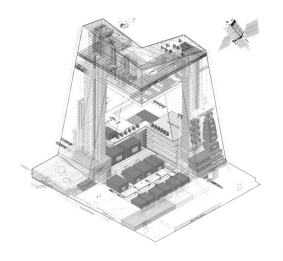

1 CCTV 新总部大楼的环形组织结构
2 RMJM 设计的阿布扎比资本中心塔

体系不仅能创造和谐的建筑形式，在某种程度上也能成为形式表现的主导因素，建筑师可以利用结构选型与细部构造创造出富有表现力的新建筑形态（图4、图5）。

随着结构技术的发展，很多建筑师已不再满足于对结构构件的象征性表现，而是将结构体系作为有机整体，使其成为形式中最富表现力的因素，创造悬挑、扭转、透空等极富视觉冲击力的造型。近期库哈斯领衔的大都会建筑事务所在新加坡设计了第一个委托项目——36层"悬浮"式公寓住宅，该建筑占地6 100 m²，建筑面积2万 m²，总建筑高度153 m。由于建筑位于城市中心区黄金地段，根据相关法规要求必须在开发过程中提供相应的公共活动用地。在这里，建筑师创造性地采用悬挑结构，四个相对独立的塔楼围绕中心十字型钢梁柱布置，结合各方向的景观视线形成不同的高度。每座塔楼底部都有从十字型钢柱伸出的钢质托盘作为支撑，同时每层楼板也有从结构中心悬挑出的钢质桁架支撑，最终形成悬挑的建筑形式。建筑师对于结构体系的创造性选择与运用，不但解决了用地的限制问题，也使结构逻辑成为形式表现的重点。

二、形式表现的主要趋向

高层建筑不仅是时代发展的产物，还是与时代紧密联系的建筑类型。新世纪的社会变迁、经济发展、技术进步均对高层建筑形式产生直接而深刻的影响。人们的生活方式、审美取向以及价值观念的变化在某种程度上引领高层建筑的发展。

1. 关注表里共生的内在逻辑

高层建筑诞生于19世纪末期，它紧跟世界建筑的发展脚步，历经古典复兴、现代主义、后现代主义、晚期现代主义等多个历史时期。现代主义提出"形式追随功能"的口号，强调建筑功能的重要性和主导地位。后现代主义者则强调建筑应表达一定的意义，传递对历史和环境的尊重。但无论是现代主义还是后现代主义，两者都没能辩证地看待建筑物质与精神的双重属性。当代建筑思潮与理念的多元化已蔚然成风，在高层建筑领域，高技生态主义、新折中主义、新地方主义等诸多流派竞相登场，充分表现着各自的设计理念与建筑特色。尽管各种流派的思想观念不尽相同，但它们越来越关注形式作为实现功能与传递意义间的工具作用，在建筑创作的认识上也把形式推到关联功能与意义的中心地位。尤其在社会、经济和技术飞速发展的当代，高层建筑内部功能急剧扩展，几乎涵盖商业、居住、办公、休闲、教育、医疗等大部分建筑类型，高层建筑的象征意义也急剧扩充，不仅要体现场所环境的内涵，更要成为时代精神的象征。

高层建筑的内部功能组织、结构构造等系统必须是优化、高效的，这也是经济性的体现；同时，它必须在城市环境中传递更多的信息，表达更丰富的象征意义，这也是艺术性的集中体现。因此，关注表里共生的内在逻辑，这已经成为新世纪的高层建筑形式表现的主要趋向之一（图6）。

2. 关注审美主体的体验感受

审美活动是人类认识活动的一种，是审美主体对外部世界的感受与反映。当作为客体的建筑带给主体的感受与人内在心理结构相一致时，客体与主体间就会建立某种联系，这种同构使审美主体在情感上与客体产生共鸣，这就是建筑审美活动的发生。我们从中不难看出，审美主体发挥了很大的能动作用。建筑为人提供居住和活动的场所，是生活模式的物化，这是建筑物质属性的体现。但从古至今建筑的内涵与意义远不止于此，人们总是在获得一定物质保障的同时转而追求更高的精神层面。尽管这种追求在一定时期呈现为一种有意识或无意识的行为，但这种追求从来没有停止过。

新世纪人们生活方式和价值观念的多元化发展，必然导致更多建筑需求的提出。这些需求既有物质上的也有精神上的，而这种满足恰恰是通过人们对建筑身临其境的体验与感受获得的。建筑的空间体验主要表现为人们对建筑内部空间、材料形式、外部形态以及人心理情感的直观而综合的反应。因此关注审美主体的体验感受，也是新世纪高层建筑形式表现的主要趋向之一。

3. 关注象征意义的视觉传达

高层建筑的象征意义包含两方面：一方面是高层建筑作为时代精

3　MAD设计的加拿大梦露大厦
4　RMJM设计的奥克塔摩天楼
5　卡拉特拉瓦设计的瑞典马尔默扭转大厦

神的象征，另一方面是高层建筑表达场所环境的意义，时代精神与场所环境共同搭建起一个时空坐标，使高层建筑的象征意义得以展现。时至今日，高层建筑业已成为现代城市的重要组成部分，但是它的发展和普及丝毫没有削弱其象征意义的表达，它依然是我们所处时代的最好见证之一。

在某种意义上来说，建筑艺术是一种象征艺术。由于人们对于建筑艺术的感知大多是一种无意识行为，因此我们需要关注象征意义在视觉传达过程中的方式，使之能够更好地、更准确地被人理解。"对建筑艺术而言，如果建筑师想使他的作品达到预期效果，并不致因译码的变化而被糟蹋，他就必须利用许多流行的符号和隐喻所具有的余度，使建筑有更多的代码性。这一原则同样适用于高层建筑，因为强有力的隐喻水准使得人们不得不在无意识中正视形式语汇的言外之意，从而在我们译读建筑时产生与自己的生活或传统更为接近的情感。"在传媒手段发达的当代社会，视觉符号已逐步取代语言符号成为文化象征的主要载体。因此关注象征意义的视觉传达，成为新世纪高层建筑形式表现的主要趋向之一。

三、形式表现的核心本质

当代不断发展的建筑技术为形式表现带来更多的自由度。在高层建筑领域，随框架体系、框筒体系的日益成熟，内部承重结构与外部维护结构越来越呈现出一种分离特征，建筑表皮因而能相对独立地存在。这种自由度与独立性为高层建筑形式表现的多样性奠定基础，在缔造丰富视觉感受的同时也产生很多非理性的、含混的甚至怪诞的形式。对高层建筑而言，一方面结构体系和材料组织是形式表现的根本，另一方面经济技术、环境文脉、社会文化等又要求高层建筑具备相应的美学价值，因此新世纪高层建筑形式表现的核心本质就体现在表皮的理性建构与形式的美学价值方面。

1. 表皮的理性建构

纵观世界建筑的发展历史可以看出，无论是以石材为主的西方古典砌块建筑还是以木材为主的中国传统木构架建筑，其形式都具备一种自主的发展状态，即形式跟随自身某种逻辑的产生、成熟乃至衰落。古典建筑形式所遵循的逻辑就是对材料特性的发掘及对材料组织关系的表达，如石材以拱的形式出现，木材以构架的形式出现等。现代建筑以工业化建造技术为基础，大量使用混凝土、钢材、玻璃、金属板等现代材料，从某种意义说，正是这些材料以及它们的组织构造方式使现代建筑获得了生命力。当代高层建筑亦然，无论是钢筋混凝土构筑的框架承重结构还是玻璃幕墙围合的维护结构，材料的特性以及材料的组织结合方式带来的表现性特征已成为高层建筑形式塑造的根本（图7，图8）。

新世纪高层建筑日益成熟的结构体系为造型的灵活性提供基础与保障，同时这种成熟又导致建筑内与外的进一步分离，使高层建筑的外部维护结构成为可相对独立存在的表皮，这种特性既为形式表现的丰富性带来契机，却也可能导致各种肤浅装饰风格的拼贴运用，导致建筑品质的缺失。因此，笔者认为建筑表皮的理性建构应该具备以下两点内涵。

一方面，建构应是真实的。尽管现代技术赋予高层建筑表皮一定的独立性，但表皮仍是整体建筑中有机的组成部分，是高层建筑内在的功能空间、结构体系、象征意义等生成逻辑的理性反映，是表里共生的产物。这种真实还体现在对材料的理解和使用上，如通过石材来营造庄严的、纪念性的氛围，通过玻璃幕墙、金属板来营造轻盈的、虚实相映的气质。因而，表皮建构的这种真实性能在一定程度上避免因形式生成的逻辑混乱或内涵缺失导致的"形式主义"及过度"装饰化"。另一方面，表皮的理性建构应是诗意的。弗兰普顿在《建构文化研究》中将建构称为诗意的建造，他认为建构包含技术问题，但又绝不仅是一个建造技术的问题。从某种角度来看，表皮的建构过程可看作是建筑师对各种材料的组织过程，但这个过程不是机械的，而是在组合调配材料的过程中融入建筑师的情感与创造，也正是在这一过程中材料超脱自然获得新的内涵。

6　OMA 设计的 CCTV 新总部大楼
7　福斯特及合伙人事务所设计的伦敦瑞士再保险公司总部大楼
8　让·努维尔事务所设计的巴塞罗那 Torre Agbar 大厦

2．形式的美学价值

美学是研究审美主体与审美客体之间相互关系与本质规律的学科，它涉及哲学、心理学、艺术学等诸多学科。《现代汉语词典》对价值这个概念的解释为"积极作用"，李德顺先生认为："价值这个概念所肯定的内容，是指客体的存在、作用以及它们的变化对于一定主体需要及其发展的某种适合、接近或一致。"由此可见，美学价值并不是审美客体的一种基本属性，而是审美客体对审美主体的作用和效果，是审美客体和主体间共同参与形成的一种特定的动态关系。就高层建筑而言，其形式表现的美学价值就在于建筑形式能否对人的各种物质需求与心理情感产生积极的作用与功效。

高层建筑的形式表现很容易通过视觉和知觉的手段被人感知，但其形式的美学价值并不是高层建筑自身的一个内在属性，也不是完全由审美主体决定的，而是产生于高层建筑形式与人之间的相互作用以及两者间的有效关系。这种关系是复杂、可变的，作为审美主体的人在这种关系中往往有很大的能动作用，如不同的人或同一个人在不同时间会对同一建筑形式产生完全不同的感受。因而，明确形式的美学价值内涵与生成机制，能够让我们在价值观念多元化的今天更好地把握高层建筑形式表现的核心本质。

结语

综上所述，以21世纪为时代背景，探讨了新世纪高层建筑形式表现的三个基本特点：功能空间的呈现、环境意义的展现以及结构逻辑的显现。分析了形式表现的三个主要趋向：关注表里共生的内在逻辑、关注审美主体的体验感受以及关注象征意义的视觉传达。同时，结合建构、美学、价值等理论，提出了表皮的理性建构与建筑形式的美学价值是建筑形式表现的核心本质，进而，对新世纪高层建筑形式表现特征进行理性解析。

参考文献

[1] 周剑云．自主的建筑形式——简介《建筑形式的逻辑概念》[J]．世界建筑，2003（12）：80-81．
[2] 梅洪元．中国高层建筑创作理论研究 [D]．哈尔滨：哈尔滨工业大学建筑学院．2000．
[3] 李德顺．价值论 [M]．北京：北京人民大学出版社，1987．
[4] 庄惟敏．几个观点、几种态度、几点呼吁 [J]．建筑学报，2004（01）：68-69．
[5] 陈伯冲．建筑形式与图像语言 [J]．建筑师，1995（12）：32-42．
[6] 戴维·B·布朗宁，戴维·G·德龙，路易斯·I·康．在建筑的王国中 [M]．马琴，译．北京：中国建筑工业出版社，2004：204．
[7] 李祖原建筑事务所．台北101大楼设计理念 [J]．时代建筑，2004（04）：89-91．
[8] 马进，杨靖．当代建筑构造的建构解析 [M]．南京：东南大学出版社，2005：143-150．
[9] 徐千里．重建全球化语境下的地域性建筑文化 [J]．城市建筑，2007（06）：10-12．
[10] 华黎．雷蒙德·亚伯拉罕访谈录 [J]．世界建筑，2003（04）：108-109．
[11] 金秋野．库哈斯方法：当建筑学成为反讽批评 [J]．建筑师，2006（06）：54-58．

作者简介

梅洪元　全国工程勘察设计大师
　　　　哈尔滨工业大学建筑学院院长、教授、博士生导师
　　　　哈尔滨工业大学建筑设计研究院院长、总建筑师
　　　　《城市建筑》主编

李少琨　哈尔滨工业大学建筑设计研究院研究生

FROM THE ORTHOGONAL TO THE IRREGULAR: THE ROLE OF INNOVATION IN FORM OF HIGH-RISE BUILDINGS

从正交四方到随意变换
——设计创新在高层建筑形式中的角色

菲利浦·欧德菲尔德 安东尼·伍德 | Philip Oldfield Antony Wood

自从高层建筑在19世纪的北美诞生以来，其形式就一直相对稳定，并且通常是根据高效的楼层平面在垂直方向上的累积衍生而来的。建筑中的商业利益造就了直线围合的玻璃以及钢筋的方盒子，一时间主宰了全世界的城市景观。创新常常被局限于一个精雕细琢的入口，一个装饰性的华盖或者是尖顶，而在这两者中间却少有作为。或许唯一对此离经叛道的行为就是战前北美在1916年出台的纽约城市区划法规。它要求建筑后退，使摩天大楼下的街道能够接受到阳光和空气。它造就了我们熟悉的"婚礼蛋糕"式的摩天大楼，这种建筑形式一时成为天际线的主导（Oldfield et al, 2009）。然而，即便是那一时期大放异彩的标志性建筑，诸如克莱斯勒大厦（1930）以及帝国大厦（1931），在形式和室内空间上也与商业模型并无大异。

近年来高层建筑的形式有了突飞猛进的变化，建筑的范式由"四方正交"变成了"自由变换"。新的设计分析工具与建筑材料、结构系统把设计师从方方正正的形式桎梏中解脱出来，使那些看似不可为的外形和不寻常的形式出现在全球天际线中。这些与众不同的不规则高层建筑被世界高层都市建筑学会（CTBUH）所关注，并将其2006年在芝加哥召开的国际会议主题定为"跳出方盒子思维：锥形、倾斜、扭曲的高层建筑"（Wood, 2007）。

这种变革的潮流在北京CCTV大楼上展现无疑，其设计方OMA将电视制作的全过程结合在一个互动的环状空间中，这个空间由四个要素构成：一个9层的"基础"，两座分别向内倾斜6°的塔楼，一个9～13层的悬挑结构——它悬浮在36层高的空中（Carroll et al, 2008）。这是一个革命性的设计，不仅仅表现为反重力的形式，而且表现为结构和功能的创新（图1）。

我们也可以说，高层建筑的原型已经被发展到了极致，它似乎为设计师提供了无限种可能。但是当我们进行更细致的观察时就会发现，很多不规则且形式复杂的高层建筑仅仅是为了成为一个不同凡响的城市"雕塑"而设计的，与当地的气候、文化、氛围环境并没有太大的关系。高层建筑的发展确实在进步，但是很多不规则、不寻常的高层建筑形式仅仅是为了成为该城市在国际舞台上崭露头角的标志而已。之前，高度突破是标志性建筑在城市上空傲然挺立的原动力，现今，形式在高层建筑设计中已经成为与高度同等重要（如果不是更为重要）的因素，而世界各个城市以及各地的开发商正在绞尽脑汁获得属于他们自己的扭曲、倾斜或者说特征鲜明的高层建筑。

"世界上的一些城市规划师被委以重任，要保证他们自己的城市在争夺国际关注的竞争中成为佼佼者，而占据这些规划师思维的建筑形式正是摩天大厦。"（Kong, 2007）

对高层建筑标志的需求造成了这样一种病态的现象：高层建筑正在成为一种孤立的建筑形式，茕茕孑立，不用因地制宜地设计，随时可以被照搬到世界上任何城市。是什么决定了高层建筑出现的时间和地点，使它们免于成为席卷全球的所谓"国际"高层建筑单一文化的一部分，并且还能够通过它自身体现当地的文化？高层建筑显然不属于能与周围环境"高度融合"的类型，它注定要耸立云端，要主导它周围的环境。但这并不意味着它不能成为城市构成中的积极因素。它可以也应该像高质量的低层建筑那样积极与周围环境互动，在环境和基地中找到自己的定位并努力成为视觉标志。

"在建筑历史中我们现在的位置绝不仅仅是象征主义和标志性。除了象征之外，建筑的特定环境、文化、生活方式以及我们使用的建设工具和方法都应该是为新型建筑增加价值甚至是为城市转型做出贡献的基础。"（Gang, 2008）

不幸的是，现在看来，尽管我们有许许多多的机会去创造新的形式来推动高层建筑原型的发展，但在更多的情况下我们还是在为形式而形式。

高层建筑形式创新的机会

显然，在创造不规则而与众不同的高层建筑形式方面还存在诸多

1 CCTV 大楼,北京(Copyright by Arup)
2 至点大厦设计因循芝加哥的太阳轨迹做出直接反应（Copyright by Gang Architects）
3 迪拜 Burj Khalifa 大厦的形体与造型设计考虑到高空风荷载（Copyright by Skidmore Owings & Merrill LLP）

挑战（最主要的是结构和建造难题，但毋庸置疑的是，这种形式具备使高层建筑原型脱胎换骨的潜能并且会对全世界各个城市产生积极作用。下面简述未来高层建筑形式创新的三种可能。这些机会为新的高层建筑形式脱离规矩的方盒子提供了起点，同时也使高层建筑超越了一般意义的雕塑性。它们展示了高层建筑形式如何才能更好地与基地环境相呼应，达到更令人叹为观止的高度并且在这样的高度容纳城市重要的新功能。

1. 环境和可持续性

纵观当代世界环境格局，全球气候变化可算是现代世界面临的最大挑战，而人工建造环境是温室气体排放的罪魁祸首之一。建筑占全球主要能源消耗的30%~40%（UNEP, 2007）。在这样的背景下，国际社会对"高层建筑是目前和未来城市中合理建筑原型"的说法褒贬不一。有些人相信高层建筑使高度集中的人口与其自身尺度带来的节约相结合（因此节省了交通开支以及城市和郊区的扩张），由此高层建筑也就成为可持续发展的不二选择。另一些人认为高空作业潜在的能源消耗以及高层建筑对整个城市区域的影响注定使它对环境产生消极作用（Roaf et al, 2005）。

高层建筑的形式创新因此应当在提升高层建筑环境表现上起到重要作用。具体来讲，高层建筑应更好地对当地气候做出反应，在设计建筑形式、根据日照和风力情况确定朝向方面采用有针对性的办法，如降低不必要的太阳能吸收并鼓励被动通风。

"作为当地最具地域特征的因素，气候为设计师因地制宜的建筑表达提供了最合理的起点，因为气候是决定当地居民生活方式和地景生态的主导因素之一。"（Yeang, 1996）

许多高层建筑还尚未对气候做出足够的反应，幸运的是有实例表明越来越多的建筑正在以"对环境做出恰当的反应"作为形式设计的主要出发点。最早的例子之一是由SOM设计、1984年竣工的国家商业银行。这座建筑位于酷热难耐的沙特阿拉伯城市吉达，设计将玻璃幕墙凹陷进建筑内部使其不接受太阳直射，遮挡其外的空中花园位置经过精心设计，切入三角形平面的每一面。之后，相似的形式于1997年被诺曼·福斯特用在了法兰克福的商业银行大厦中，这一次这种形式达到了更好的效果。1992年，梅那拉·梅西加尼亚大楼建成于吉隆坡，其建筑形式随高度变化而不同以呼应当地气候。该设计中，电梯井被挪到建筑的外沿，为内部遮阳。与之相对的，削弱建筑体块得到的空中花园承载着连绵的景观和植被。最近的一例是Studio Gang的至点大厦建筑方案，它对芝加哥的太阳轨迹做出直接反应。南立面的锯齿状形体为住户遮挡住夏日毒辣的太阳，减少了太阳能摄取；冬天，在太阳轨迹较低的时候又为室内提供被动式太阳能供暖（图2）。

高层建筑的形式创新同样可以表现在基地现场使用低碳排甚至是零碳排的能量生产来源。以巴林世贸中心（麦纳麦）、珠江大厦（广州）、阶层大厦（伦敦）为例，它们都利用涡轮机增加风速，以得到清洁的能源。我们不能否认，它们的建筑形式也同样激动人心。

2. 风力及结构工程

在减小建筑承受的风荷载并增加结构效率的手法中也存在着创新的机会。这种手法可以在增加建筑高度的同时减少材料的使用。过去和现在研究的重点都在于不同建筑形式对风荷载的影响以及建筑的运动状况（Denoon, 2006; Irwin et al, 2008）。相对于四方正交和对称的形式，不规则的建筑形式在这方面似乎更具优势。尽管在结构骨架设计方面不规则形式会对结构设计提出挑战，但是它在减小风荷载及建筑应对风荷载能力方面却有极大的帮助（Ali & Moon, 2007）。

对于高层建筑（特别是那些达到惊人高度的建筑），风洞测试是设计过程中必不可少的一部分。建筑形式和形状上很小的改变都会对结构和风荷载产生重大的影响，从而影响到建筑材料的数量和成本。全世界几栋最高的建筑设计过程中，生成最合理的建筑形式在相当的高度上抵抗巨大的风荷载显得至关重要。目前世界最高的建筑是Burj Khalifa，它828 m的高度令人瞠目结舌（图3）。而在设计过程中，该

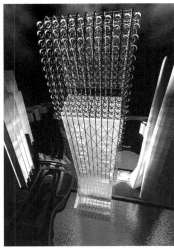

建筑的形式和几何构成更是经历了数轮风洞测试才最终确定。这一过程的结果是通过促使被打乱的漩涡顺着建筑的形体流下，进而对施加在该建筑上的风荷载进行"误导"，从而大大地减小施加在建筑上的风荷载。

根据风荷载对高层建筑形体进行最优化设计，对环境与经济也大有好处。上海中心大厦632 m高的不对称结构由Gensler设计，它能够降低24%的风荷载，相应地能够节省32%的建筑材料（Gensler, 2009）。这样不仅能够减少5 800万美金的建设成本，而且可以节约上千吨钢材和混凝土。由于施工和材料通常会占据整个建筑生命全周期的20%左右（Kestner, 2009），这种节省将对降低大楼整体碳排放具有极大作用。

3．连接及联通的高塔

高层建筑的另一个创新形式的潮流应该就是将高层建筑相互连接了。在未来十年中城市人口将不断增长，将占地面积颇大的"水平"建筑结合整理，收纳在垂直的高层建筑中将成为趋势。这一举动是对传统的办公、住宅和酒店功能的一大挑战，同时也是对目前由这些功能所主宰的高层建筑的挑战。对于设计师而言，这种挑战就是要开发出新的建筑，将公园、运动场所、休闲设施、学校、农田以及其他社会交流功能融入其中，把城市生活的丰富多彩和生动活力带上云端。一种可能的解决方案就是将几栋高层建筑连接起来，从而在一定高度形成较大的空间，或者可以由连接几栋高层建筑的天桥来实现这些水平功能。

如果我们把垂直的高层建筑仅有的连接设在地面层或者是地下，这就没有什么值得称道的了。将几座高塔在空中连接可以有效地缓解高层大楼作为孤立建筑存在的僵局，增加城市的丰富性并使天空有机会成为公共领域。此外，9·11带给我们一个警示：在高处将大厦联通可以通过增加水平疏散区域提高建筑的安全性。一旦建筑出现危险，能够将住户在短时间疏散到除地面层之外的安全楼面上显得非常合理，特别是在高层建筑的垂直疏散通道被切断而无法到达地面层的紧急状况下。

尽管这种设想听起来有些异想天开，但是它已经在全世界的一些建筑中得以实行。ARC Studio在新加坡设计的Pinnacle @ Duxton由7座高塔组成，26层与50层有宽大开敞的空中花园将高塔串联起来。空中花园为住户和公众提供了良好的绿色景观，并在增加安全系数的同时给城市景观锦上添花（图4）。同样是在新加坡，由Moshe Safdie和Aedas共同设计的Marina Bay Sands由三座徐缓起伏的塔楼组成，在它们的顶层之上冠以340 m长的空中花园。在这两个案例中，天桥都是其使用的建筑语汇、建筑形式与住户体验的重要元素。

尚处在方案阶段的高层建筑联通项目一般更加极端。在世贸中心大厦重建竞赛过程中，很多由下曼哈顿发展公司、纽约及新泽西港务局所支持的项目（Stephens, 2004）就采用了在高空联通塔楼的手法。例如，United Architects的设计提案中5栋水晶般的高塔相互倚靠并在340 m处融合，创造出一座大教堂式的建筑（图5）。在它们碰撞位置，塔楼为公共空间提供了5个连续的楼面。此外，它们通过更多的疏散路径和多水压源的喷淋系统增强了大楼的安全性（Wood & Oldfield, 2007）。

这些建成的与提案中的建筑或许为我们提供了联通城市的愿景：更多的新型建筑在高空之上为我们创造出一片天地。

未来的设想

以下简述6种未来高层建筑的设想，它们的创作者都是作者指导的建筑研究课题的学生。前面三个作品由诺丁汉大学的学生设计，

4　新加坡 The Pinnacle @ Duxton（Copyright by ARC Studio）
5　纽约世贸中心重建方案采用五栋塔楼融合的形式，为公共设施提供大而连续的楼面（Copyright by United Architects）
6　都市风电厂设计方案（Copyright by Adam Chambers 与 Alex Dale-Jones / 诺丁汉大学）
7　都市风电厂通透且轻盈的体量（Copyright by Adam Chambers 与 Alex Dale-Jones / 诺丁汉大学）
8　层叠庭院建筑全景图（Copyright by Minh Ngoc Phan / 诺丁汉大学）

9 层叠庭院内部景观（Copyright by Minh Ngoc Phan / 诺丁汉大学）
10 渲染我的泰晤士建筑中某垂直聚落的剖切透视图（Copyright by Chandni Chadha, Arham Daoudi, Elnaz Eidinejad 与 Akshay Sethi / 诺丁汉大学）
11 渲染我的泰晤士建筑全景图（Copyright by Chandni Chadha, Arham Daoudi, Elnaz Eidinejad 与 Akshay Sethi / 诺丁汉大学）
12 海勒大厦建筑全景（Copyright by Kent Hoffman 与 Mark Swingler / 伊利诺伊理工大学与 CTBUH）

是可持续高层建筑研究生课程的一部分。该课程由 Philip Oldfield 和 David Nicholson-Cole 任教。后面三个作品的设计者来自由 Antony Wood 指导的伊利诺伊理工大学建筑设计专题。这些项目都很好地向我们展示了创新的功能形式是如何与一个地区在实体、环境以及文化特征上相关联的。

1. 都市风电厂

这座为伦敦金丝雀码头设计的建筑，其倾斜的形体主要是考虑环境因素。其主要理念是让建筑整体向南倾斜以便为它自己提供遮阳措施。建筑倾斜的角度根据环境进行最优化选择，在夏天太阳角最大的时候避免不必要的太阳能吸收；冬天，当太阳角较小的时候，遮阳减少，从而得到更多的被动式太阳能供暖。结构上，建筑的倾斜通过三角形的钢制斜肋构架达成。在建筑的东西立面建有垂直的混凝土墙以及剪力墙（图6，图7）。

除了通过自身遮阳措施减小能耗之外，这一设计还试图在基地现场利用可再生能源满足一定的建筑能耗需求。在大楼的顶端和四角（也就是风速最快的地方）建设了风电场。它是将大量的复合扩张涡轮机（CATT）陈列在轻质张力骨架上形成的结构。其设计成果深植于当地环境中，与环境特征紧密结合，同时为在高空进行可持续设计提供了大胆的假设。

2. 层叠庭院

庭院围合式的建筑是中东地区重要的民居形式。它为住户提供了半开放式的庭院以躲避沙漠毒辣的太阳和疾风，与此同时也在一定程度上保护居民的隐私。这个项目旨在利用高层建筑形式重新对围合庭院进行诠释（图8，图9）。

建筑四四方方的外观为内部创新有趣的空间蒙上了一层神秘的面纱。建筑由一系列六层高的聚落叠摞起来，每一个聚落中心都围绕着一个半公共的庭院。庭院的设计是为了应对高层建筑缺少娱乐交流空间的弊病，高层建筑也正是因此才不适合儿童和家庭居住。在该建筑中，庭院相当于垂直聚落的社交中心并且也为提升安全性、景观和私密性做出了贡献。随着建筑高度的增加，庭院渐渐转向北面面向滨海大道的最佳景观。巨大的雕刻屏为这些空间提供遮蔽，同时允许波斯湾上的丝丝微风通过屏风传到室内。庭院周围住宅设计的灵感也是来源于传统建筑形式：它们中有很多都是为大家庭设计的，不同的房间可以通向同一个私有庭院空间。外部建筑立面的设计更强调内部庭院，而玻璃窗受限于连续但狭窄的开槽，造成建筑不透明的外观。

3. 渲染我的泰晤士

本设计试图根据色彩和艺术类型来整合高层社区。该建筑15层以下是一个设计学院，住宅楼位于其上。与"层叠庭院"相似的是，这座高层建筑被划分成了一系列垂直聚落，每个聚落共享一个空中花园和社交空间，例如幼儿园、图书馆和餐厅。这个设计拒绝使用高层建筑固有的庞大整体形体，而更重视用聚落和空中花园对高层建筑进行片段式的表达。这种手法使得建筑更像是一系列颇具趣味的低层建筑的叠摞，而不是一个巨大的单一体块趾高气扬地主导它周围的环境。这成为该建筑与周围低层建筑环境之间的纽带（图10，图11）。

设计尤以建筑表皮为重点，没有或蓝或灰的典型玻璃幕墙，取而代之的是将色彩和活力按照富有意义的方式与大厦融合，而不简单是为了美观。每个立面上的色彩选择与分布都是与周围环境相辅相成的。因此南立面的主要色彩就取自旁边的集装箱城市、蓝灰相间的泰晤士河，白色的O2穹顶以及更南面以红、绿、棕三色为主的居民区。通过这种方式，每个立面都是周围环境的抽象体现。木质的百叶不仅为建筑住户提供遮蔽和私密性，同时也使立面外观更加动感多变：百叶的每一次开关都会使色彩体块的呈现产生变化。当夜幕降临，设计

学院大楼就会像空白画布一样用来放映电影、学生作品和数码展示，使建筑体验更加生动丰富。

4. "海勒大厦"——垂直健身中心

这项企划是针对孟买人口高度密集而城市缺乏休闲设施而建设的垂直健身设施与宾馆。本项目的出发点是许多运动需要大量的水平空间，而一般并不会建设在高层建筑中。该设计用以应对这一问题的方法是用一系列室内竞技场（伴有一定面积的室外草坪）来容纳各种运动。竞技场水平设在六座高塔间不同的高度上。其中两座塔楼垂直矗立，主要负责这个项目的垂直交通。另外四座在空间上成扇形打开，由此后面的塔楼不会被前面的所遮挡，从而使所有的酒店客房都能欣赏到海景。在结构方面，提供运动场地的"桥梁"也将两栋向外倾斜的建筑紧紧拴在一起（图12）。

该建筑的意图是在不同时期四座塔楼中以及各个运动场地上举办不同的体育活动，例如接待"大卫·贝克汉姆足球学校"或者承办"北印度柔道锦标赛"等等。届时各种活动将在无数个层面上同时进行，而不同楼层的酒店客房、天桥以及平台将成为这些运动的训练场和竞技场。基地层被抬升以用作停车场，与周围一圈公共/商业设施相融合，并在裙房层为板球和足球建设了很大的运动场地。一条贯穿整个建筑群的通路将所有的运动层面在水平和垂直两个方向上相连接。体量巨大的攀岩壁从主厅上方挑出贯穿几座高塔。本项目意在将大量的运动爱好者直接吸引到城市中心，并形成相当规模。它不仅要吸引居住在附近的居民，而且包括所有的都市人，特别是年轻一代，而他们当中可能会有人成为未来的麦克·菲尔普斯或是伊姆兰·卡恩。

5. "加拉沙大厦"——水库大楼

孟买的日缺水量已达到2 500万升，然而在每年季风雨季这座城市又会遭遇严重的洪涝灾害。这种极端对立的情况就成为了本项目设计的出发点。在三面环海的城市当中，情况只会更糟——当季风雨量攀升的时候，海平面会随气候的变化而上升，然而提供清洁水源的基础设施却跟不上城市人口增长的步伐（图13，图14）。

加拉沙大厦的主要功能在于用水收集、净化以及储存。基地开挖后因其地势较低而被当作"水槽"。当季风雨季来临，这个"水槽"可以收集大量雨水（基地边缘作为过滤器）。随后，收集的雨水会用水泵抽入大楼内部，经过过滤储存在这个垂直水库里。该水库为占大厦一半体量的住宅区供水。这个住宅区以阶梯状的形体出现，每个"台阶"都是一个3层的小聚落，而每个聚落共享一个设置在临近聚落屋顶上的公共花园。这些住宅体块从主要的垂直结构中悬挑出来，并且有效利用了设置在垂直结构另一面的水库的宽度。水库中的水是通过"生态机械"系统进行净化的：作为公共花园的水库屋顶通过电梯的"生态走廊"与各个楼层以及每层的花园和地面层连接起来，帮助各物种在这栋楼中移动（包括植物和动物）。该方案具有清晰的公共面，所有的休闲花园都朝向城市一侧的海面；在其私密面，所有的社区花园都朝向了另一侧的海面；地面层的水库同样被用作社区休闲娱乐的资源，居民可在水库中举行划艇及游船等活动。更重要的是，当这一地区遭受阶段性洪水袭击时，这里可以变为灾难急救中心，所有的皮划艇能够在受灾地区进行紧急救援。

6. 天空体

本项目寻求在一座高层建筑内为芝加哥几所主要大学提供学生宿舍的解决方案。这座塔楼的形体强调一个观点：高层建筑不仅仅跟它所在的基地直接相关，并且与在视野范围内的其他成百上千的基地都有着微妙的联系。正因如此，这个方案是由众多多层住宅"盒子"叠摞形成的，它们围绕着建筑中心旋转，每一个盒子都面向一个城市景观，或者是某所大学，或者是城市的地标。这座高层建筑的结构设计

13 加拉沙大厦实体模型（Copyright by Bojana Martinich 与 Teodora Vasilev / 伊利诺伊理工大学与CTBUH）

14 加拉沙大厦建筑全景（Copyright by Bojana Martinich 与 Teodora Vasilev / 伊利诺伊理工大学与CTBUH）

15 天空体建筑仰视图（Copyright by Prairna Gupta / 伊利诺伊理工大学与CTBUH）

是将混凝土核心与巨大的立柱结合，这些立柱在大楼表面曲折盘桓贯穿整座建筑，钢制桁架与斜撑保证侧向稳定。这显然是一项大胆而野心勃勃的举动，但是它也确实创造了地标性的建筑，并且它的建筑形式是深深植根于周围的城市环境特征的（图15）。

建筑的能源策略包括向每一个学生住户发放一台混合动力的电动自行车，他们可以骑车上学或者在城市中活动。自行车在骑行过程中会发电并将其储存。一天结束后，这些自行车会被插回大楼内部向建筑输送电力，弥补电力消耗。

鸣谢
本文作者在此对文中详细介绍的建筑方案作者即建筑研究课题学生表示感谢：
·诺丁汉大学
都市风电场：Adam Chambers and Alex Dale-Jones
层叠庭院：Minh Ngoc Phan
渲染我的泰晤士：Chandni Chadha, Arham Daoudi, Elnaz Eidinejad and Akshay Sethi
·伊利诺伊理工大学
海勒大厦：Kent Hoffman and Mark Swingler
加拉沙大厦：Bojana Martinich and Teodora Vasilev
天空体：Prairna Gupta

参考文献

[1] ALI M A, MOON K S. Structural Developments in Tall Buildings:Current Trends and Future Prospects[J].Architectural Science Review, Vol.50.3:205-223.

[2] CAROLL C, CROSS P, DUAN X, GIBBONS C, et al. Case Study: CCTV Building: Headquarters and Cultural Center[J].CTBUH Journal, Issue 3: 14-24.

[3] DENOON R. Predicting Wind Effects and Performance of the Irregular[C]. CTBUH Conference "Thinking Outside the Box:Tapered, Tilted, Twisted Towers". Chicago, 2006, October 25-26.

[4] KONG L. Cultural Icons and Urban Development in Asia: Economic Imperative,National Identity and Global City Status[J]. Political Geography, 26(2007): 383-404.

[5] GANG J. Wanted:Tall Buildings Less Iconic,More Specific[C].Proceedings of the CTBUH 8th World Congress "Tall & Green:Typology for a Sustainable Urban Future.". Dubai: 3-5 March, 2008: 496-502.

[6] GENSLER A. Shanghai Tower: Completing a Supertall Trio[C]. CTBUH Conference "Evolution of the Skyscraper:New Challenges in a World of Global Warming and Recession". Chicago: October 22-23, 2009.

[7] KESTNER D M. Sustainability: Thinking Beyond the Checklist[J]. Structure Magazine, June, 2009: 5.

[8] IRWIN P, KILPATRICK J, FRISQUE A. Friend or Foe: Wind at Height[C]. Proceedings of the CTBUH 8th World Congress "Tall & Green:Typology for a Sustainable Urban Future." Dubai: 3-5 March, 2008: 336-342.

[9] OLDFIELD P, TRABUCCO D, WOOD A. Five Energy Generations of Tall Buildings:An Historical Analysis of Energy Consumption in High-Rise Buildings[J].The Journal of Architecture, Vol.14,No.5: 591-613.

[10] ROAF S, CRICHTON D, NICHOL F. Adapting Buildings and Cities for Climate Change:A 21st Century Survival Guide[M]. Oxford: Architectural Press.

[11] SCOTT D. The Effects of Complex Geometry on Tall Towers[J]. The Structural Design of Tall and Special Buildings, 16:441-455.

[12] STEPHENS S. Imagining Ground Zero: Official and Unofficial Proposals for the World Trade Center Competition[M]. London: Thames and Hudson, 2004.

[13] Industry & Economics United Nations Environment Programme. Division of Technology. Buildings and Climate Change: Status, Challenges and Opportunities, United Nations Environment Program[M]. Nairobi: UNEP, 2007.

[14] WOOD A. Pavements in the Sky: Use of the Skybridge in Tall Buildings[J]. Architectural Research Quarterly, Vol.7,No.3 & 4: 325-333.

[15] WOOD A. Thinking Outside the Box:Tapered,Tilted,Twisted Towers[C/CD].Proceeding of the CTBUH Chicago Conference 2006. 10-DVD set. Chicago: Councilon Tall Buildings and Urban Habitat.

[16] WOOD A, OLDFIELD P. Bridging the Gap:An Analysis of Proposed Evacuation Links at Height in the World Trade Center Design Competition Entries[J]. Architectural Science Review, Vol. 50.2: 173-180.

[17] WOOD A. Green or Grey? The Aesthetics of Tall Building Sustainability[C].Proceedings of the CTBUH 8th World Congress "Tall & Green:Typology for a Sustainable Urban Future.". Dubai: 3-5 March, 2008: 194-202.

[18] YEANG K. The Skyscraper Bioclimatically Considered:A Design Primer[M]. UK: Wiley Academy.

作者简介
菲利浦·欧德菲尔德 英国诺丁汉大学建筑环境学院副教授
安东尼·伍德 美国芝加哥高层建筑与城市环境协会，芝加哥伊利诺伊理工学院建筑系教授

PARAMETRIC SKYSCRAPERS
高层建筑参数化设计

陈寿恒 | Chen Shouheng

参数化的"西萨·佩里"

自1977年西萨·佩里的第一座高层——纽约现代艺术馆大厦落成至今，西萨·佩里已经在世界各地设计和建造了超过40栋地标性的超高层建筑，其中包括吉隆坡双塔（Petrona Towers）和香港国际金融中心大厦（HKIFC）等（图1）。根据建筑评论家约瑟夫·格罗凡尼尼（Joseph Giovannini）的统计，"西萨·佩里每年设计建成的高层建筑规模相当于1.5栋纽约帝国大厦"。

作为现代派的建筑大师，西萨·佩里设计的高层建筑似乎和参数化设计并没有直接的联系，但是当我们把他的作品进行归纳和分类，可以发现其形体的几何构成具有强烈的规律性。同时，这种规律性伴随着建筑工程技术的进步发生着阶段性的变化。从这点我们可以发现他的设计方法和参数化设计提倡的以强调发掘和利用逻辑规律为主导的设计方法——基于规则的设计手法（Rule-based Design）不谋而合。以下是此次研究的成果。

西萨·佩里设计的高层建筑种类繁多，造型千变万化，但是我们可以用四种基本几何构成方式把它们进行分类：分支形（如吉隆坡双塔）、角部退台形（如香港国际金融中心）、多折面形（如费城的西拉大厦）和结构外墙形（包括竖向结构和菱形外墙结构）。如果按建构方法进行分类，我们甚至可以用两种简单的法则来定义它们：空间截面法则和空间点法则。空间截面法则是通过寻找平面截面的空间位置，并采用放样和布尔交集运算来生成建筑实体（图2）。绝大部分西萨·佩里的作品都能使用这种法则来进行设计，但是近年来我们也看到他对空间点法则的偏爱。空间点法则是通过寻找定义建筑边界的空间点的几何定位规律，并通过组合面的方法来生成建筑实体。西拉大厦（Cira Tower）就是一个典型的例子。邻近的角点之间都有明确的几何关系，比如角点15是在角点5和角点1连线的1/6位置上，角点67是在角点6和角点7连线的2/3位置上，如此类推，通过定义这些角点与其相邻两个角点的距离比例，这样我们便可以从中找到西拉大厦的几何构成规律（图3）。

由于西萨·佩里的高层建筑作品带有明晰的几何构成规律，我们可以通过数字化的手段（使用电脑编程用数字来定义形体特征）来演绎他的设计精髓。再者，我们可以结合形变规则（包括比例缩放、平移和扭转等）来生成千变万化的设计方案（图4）。图5和图6分别是通过使用空间截面法则和空间点法则，并结合形变规则生成的案例，它们既继承了原始设计的特点，又具有形变规则赋予的新几何形体特征。从这里我们看到了对于高层建筑形体进行参数化形变控制的一种具有可行性的方法。

高层建筑参数化形变

通过参数化对西萨·佩里的作品进行研究具有一定的局限性。虽然通过参数运算所生成的高层带有典型的西萨·佩里风格，但这是因为我们所归纳总结，并诠释成数字结构的逻辑规律都是源自于西萨·佩里的设计方案。然而这项研究却从侧面让我们确信：高层建筑的形体构成是可以通过简单的归纳法则进行分类，并通过布尔交集或者组合面的方法来构建。如果我们引入形变规则，那么所能得到的创新设计将是无穷无尽的。

基于这层思考，笔者在"高层参数化形变系统"开发项目中提取了参数化的西萨·佩里研究成果的精髓，并把它发展成为一套具有普遍实践意义的参数化高层形体构成系统。以下是对这项研究成果的简短介绍。

高层建筑参数化形变系统不以个体建筑项目为研究对象，而是致力于研究几何类型学，归纳出适用于高层建筑设计的几何形体。同时，它把这些可能的形体组合方式进行配对并罗列出来（图7）。最后也是最关键的一步，即我们发展了一套能够解读这些几何形体及其配对组合的数字化运算系统。这套系统的核心内容是对这些形体的拉伸、比例缩放、平移、扭转和布尔交集运算等形变方式的控制（图8），甚至可以把多个不同的形体组合搭配在一起形成群组。

利用参数化技术，我们可以生成由这些特定组合或组合群组构成的变参数设计多选方案（图9）。在这个项目中，我们归纳并总结出了50多种适用于高层设计的原始几何形体、1 400多种形体配对组合形式，以及不计其数的组群。由它们衍生出来的参数化方案将会是无穷无尽的，因而这个系统的应用面也将非常广泛。

高层建筑 | 理论研究

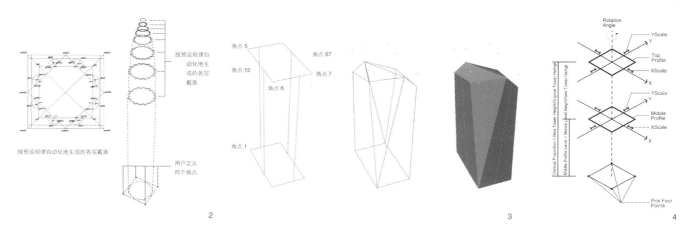

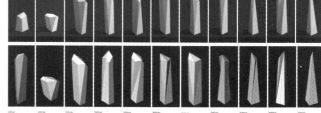

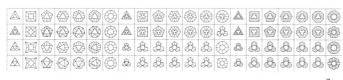

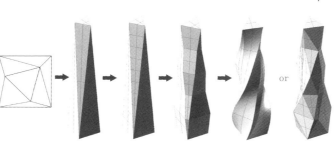

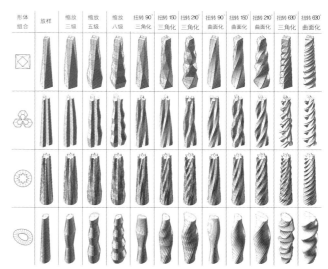

1 西萨·佩里的高层建筑作品一览（图片来源：Sections Through a Practice: Cesar Pelli & Associates）
2 空间截面法则
3 空间点法则
4 形变规则
5 空间截面法则和形变规则结合使用生成的新方案（三维打印实体模型）（图片来源：Embedding Methods for Massing and Detail Design in Computer Generated Design of Skyscrapers）
6 空间点法则和形变规则结合使用生成的新方案（三维数字模型和对应的参数设定）（图片来源：Embedding Methods for Massing and Detail Design in Computer Generated Design of Skyscrapers）
7 部分几何形体组合的范例（图片来源：Tall Building Form Generation by Parametric Design Process）
8 通过拉伸、比例缩放、旋转、Loft和三角化等形变方式生成的高层方案过程示意
9 由特定组合或组合群构成的变参数设计多选方案范例

应用"高层参数化形变系统"的实践案例

这里让我们以诺曼·福斯特（Norman Foster）、Gensler、SOM以及SHDT的方案为例来共同分享一下这套全新的"高层参数化形变系统"的潜在作用。与此同时，为了易于理解，笔者将这些方案使用"高层参数化形变系统"进行自动化模拟生成的具体操作过程列举如下（图10~图13）。

上述高层建筑方案极具时代特征，其形体复杂、富于变化且个性突出。其中Gensler设计的上海大厦已于2009年破土动工，我们坚信它的落成将给上海市的城市天际线勾画上浓重的一笔。

通过上述分析，我们发现参数化的高层建筑可以通过简单的形体构成方式来进行构建。"高层参数化形变系统"向我们展示了一套高层形体构建系统，它能够通过自定义的数字化平台自动生成规律性的形体构成，控制高层建筑形体的构建。

高层建筑参数化的细部构建

在参数化的高层设计中，我们除了要处理高度复杂的形体外，还要解决这些复杂多变的形体带来的非标准化部件的构造问题。包括罗曼·福斯特和弗兰克·盖里（Frank Gehry）在内的建筑大师都拥有自己的数字化专业技术团队为每个个案开发定制数字平台，解决实际的构造问题，这是他们的方案能够实现的技术保障。

同时，我们也能在市场上找到一些类似的专业构造和施工辅助软件，如由盖里技术（Digital Technology）开发的"数字方案"（Digital Project）和由Autodesk开发的Revit等等。这些数字平台都可以辅助扩初和施工图设计，它们对前期方案设计的帮助并不明显，这是参数化设计普遍推广仍然比较滞后的一个主要原因。

为了填补参数化设计在设计初始阶段辅助细部构建的空白，笔者开发了一套以元件化建模为基础的变参数外墙自动化三维建模系统——RHIKNOWBOT。该系统可将高层玻璃幕墙的三维建造过程自动化，因而可以用于处理高度复杂的非标准化部件。RHIKNOWBOT提供一种自定义的细部构造方法（图14）和参数化的用户界面（图15），它通过三维模型投影的方式把用户自定义的细部构造在指定的建造面上生成三维实体模型。

在设计初始阶段，这些模型可以被用于方案表现，随着设计进入扩初和施工图阶段，它们可以被用在专业的工程软件中，进行力学、材料或者声学计算。值得一提的是，RHIKNOWBOT同时提供部件的替换功能，用户可按设计意图对细部构件进行修改或替换（图16）。

自从2008年底推向市场以来，RHIKNOWBOT已经逐步得到一批前沿建筑师的青睐。图17和图18是两个由RHIKNOWBOT完成并通过三维打印机输出的高层设计方案：鼓形塔和扭转的超高层。

结语

随着电脑编程技术在建筑设计界逐步得到推广，参数化高层建筑设计也开始绽放光彩，多个极具时代特色的参数化高层建筑方案相继在世界各地出现，这些方案的成功有赖于数字化技术日新月异的发展，更要归功于建筑师对形体几何构成认知的进步。本文是对参数化高层建筑设计的一次有益探讨，希望它能引起读者对这种新设计方法广泛和深入的讨论以及对参数化技术开发的兴趣。

参考文献

[1] Chen Shouheng, etc. eds. Computational Constructs: Architectural Design, Logic and Theory[M]. Beijing, China: China Architecture & Building Press, 2009.

[2] SUZUKI HIROYUKI, BARRENECHE PAUL, GIOVANNINI, Joseph and Cesar Pelli & Associates. Sections Through a Practice: Cesar Pelli & Associates[M]. Ostfildern, Germany: Hatje Cantz Publishers, 2003.

[3] Chen Shouheng. Embedding Methods for Massing and Detail Design in Computer Generated Design of Skyscrapers[M]. Cambridge, MA, USA: The MIT Library, 2006.

[4] PARK SANGMIN. Tall Building Form Generation by Parametric Design Process[M]. Ann Arbo, USA: ProQuest Information and Learning Company, 2005.

作者简介

陈寿恒 Design and Technology（SHDT）主持设计师

高层建筑 | 理论研究

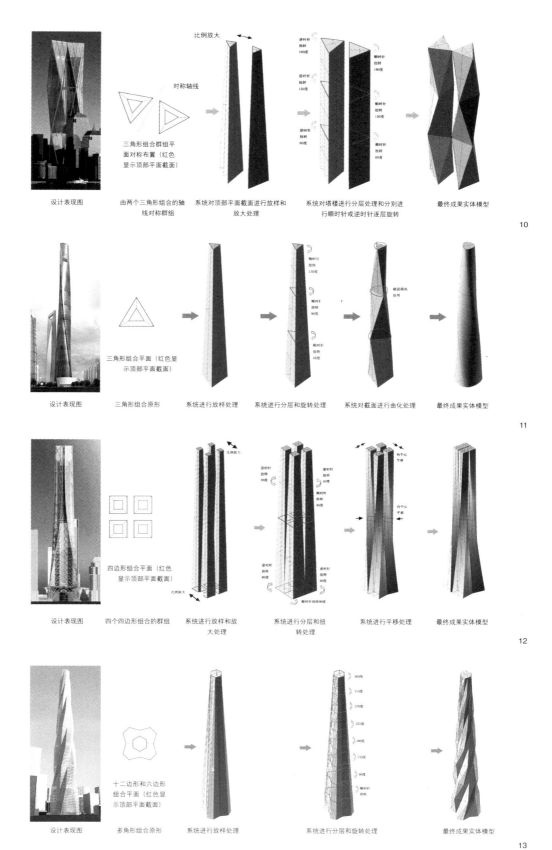

10 诺曼·福斯特设计的纽约世贸改造投标使用"高层参数化形变系统"自动化模拟生成的操作过程（设计表现图来源：http://forum.skyscraperpage.com）
11 Gensler设计的上海大厦使用"高层参数化形变系统"自动化模拟生成的操作过程（设计表现图来源：http://greenbuildings-sfk.blogspot.com/2009/11/）
12 SOM设计的Transbay大厦使用"高层参数化形变系统"自动化模拟生成的操作过程（设计表现图来源：http://www.skyscrapercity.com/）
13 SHDT设计的多伦多旋转大厦概念方案使用"高层参数化形变系统"自动化模拟生成的操作过程
14 元件构造在RHIKNOWBOT中的定义方法
15 RHIKNOWBOT用户界面
16 由特定组合或组合群组构成的变参数设计多选方案范例
17 鼓形塔及细部
18 扭转的超高层及细部

035

HIGH-RISE BUILDING | THEORETICAL RESEARCH

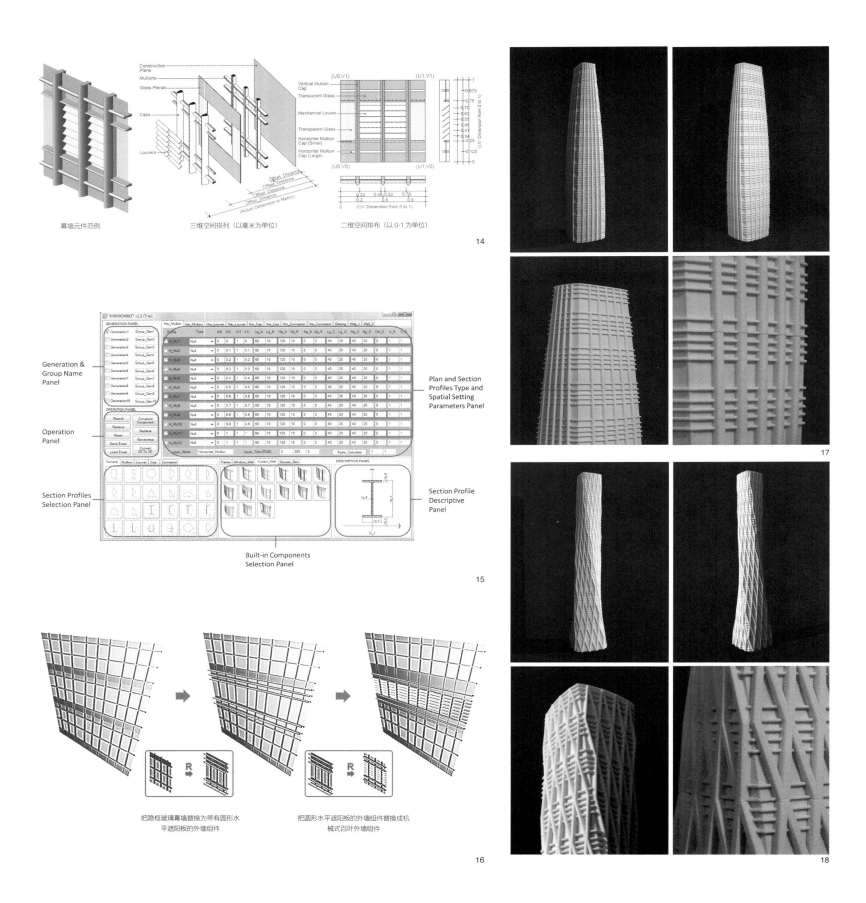

幕墙元件范例

三维空间排列（以毫米为单位）

二维空间排布（以0-1为单位）

14

Generation & Group Name Panel

Operation Panel

Section Profiles Selection Panel

Plan and Section Profiles Type and Spatial Setting Parameters Panel

Section Profile Descriptive Panel

Built-in Components Selection Panel

15

把隐框玻璃幕墙替换为带有圆形水平遮阳板的外墙组件

把圆形水平遮阳板的外墙组件替换成机械式百叶外墙组件

16

17

18

NEW PRACTICAL INTERPRETATION OF "LESS^N IS MORE"
"少即是多"理论的实践新解

山扬·李沧　安德烈斯·阿利亚斯·马德里　何塞·拉蒙·特拉莫耶雷斯 | Sangyup Lee　Andres Arias Madrid　Jose Ramon Tramoyeres

对福特主义社会神话的盲目崇拜是今日之发展迟滞不前的主要原因，现代主义思想已经完全被市场经济所取代。我们不应彻底否定现代主义，反而要恢复并深化发展现代主义思想，以便能够同社会的需求接轨。我们不仅要明确并管理好网络社会的各类需求，更应该让高层建筑的意义为世人所了解，这样新的城市和社会实体就能变得平易近人，能够引起公众的关注。我们多次提及密斯·凡·德罗（Mies van der Rohe）的"少即是多"理论，它并不是极简主义的回归，而是一种围绕有限问题的执着和坚持，从而号召突出并优化解决方案。

高层项目实践计划围绕以下五个问题展开设计工作，这些问题或直接取自现代主义设计理论，或由其理论转换而来，同时结合迪拜徽章大厦结构设计方案作以说明。

一、材料科技

网络社会要求一种全新的城市空间和建筑，必须新颖独特，富于创造力同时又兼有可持续性特点。科技的运用（包括机械和参数设计工具）将有助于我们达到既定目标，但是假如这些工具的使用完全是为了追求效率，而单纯按照功能性标准来执行的话，那么建筑本身就极有可能因为科技能够解决一切问题的错误思想而黯然失色。所以，我们的目标不应局限于将网络社会的新需要明确化、系统化，还应当同这个全新的社会实体进行交流互动，引起公众的关注。

为达到这一目标，我们回归现代主义理性的设计方式，而由现代主义大师密斯·凡·德罗设计的巴塞罗那馆正是由这一方式创造出的典范。这座4层建筑由光彩熠熠的钢板包覆，旨在展现金属工业的潜力，以一种可见方式映射出金属建筑材料最高效的利用方式。

在迪拜徽章大厦建筑方案的构思过程中，结合迪拜的文化环境是至关重要的。立面设计意在重塑伊斯兰艺术中重要的传统——通过符号而非象征来完成意义的传达（图1）。高新科技以及先进几何造型的运用令设计方案具备了以史为鉴、继往开来的意义。与此同时，立面的六边形网状结构又能够同复杂的建筑造型和结构实现系统性整合。

我们利用计算机技术，将这座伊斯兰尖顶建筑转换成能够根据不同环境进行针对性修改的直观参数向量模型。建筑立面由三个主要部分组成——南面、北面和桥，所处环境和建筑要求各不相同。

二、将楼面垂直重复（堆叠挤压）转换为自由造型（楼面拓扑变形）

现代主义超高层建筑是通过水平楼面的垂直重复构建而成的，用以填充立面的造型则以挤压堆叠的方式产生。颠倒这一过程，可先根据环境因素（日照轨迹、风量等）生成一个自由造型，然后再依照楼体造型对楼面做拓扑变形处理（图2~图4）。

例如，在迪拜徽章大厦的设计中，建筑外立面统一了整座建筑，其造型取自一个酷似虫穴的连续表面。每个表面通过一根管道连接着一对不同环境的立面，实现了压力、温度等因素的渐变，营造出表面通风的微气候（图5，图6）。为更好地适应外立面不同的环境条件，我们利用伊斯兰传统样式（伊斯兰箭头），通过参数控制的方法，根据不同条件调整开孔量并选择相应材料。

三、将分隔结构立面转换为协同结构立面

现代主义超高层建筑的突出特点就是相互分隔的结构立面，而今天的电脑技术能使两个体系协调一致。这一变革可以诠释为，从现代主义的分隔与重复的设计顺序向参数理论所倡导的系统内连续渐变以及系统间深入联系的转换。

四、将独立建筑转换为环境互动建筑

固定的墙体将现代主义超高层建筑同环境隔绝开来，不仅阻隔了内部空间与外部环境的互动，而且城市环境及公共空间也无法同建筑内部互通。当今超高层建筑设计的核心驱动力应当是创造出与环境互动的建筑，形成其自有的微气候，并重新定义建筑与城市的关系，从而为混杂而密集的功能活动营造出互动空间，增进社会交流。

这一点在我们的设计中是通过以下方式实现的。根据风量与日照轨迹等数据来创建建筑几何模型及外立面。日间，将来自海洋等风资源重新分配到室内，外立面在遮光的同时利用太阳能驱动制冷及避光设备。夜间，风自内陆吹向海洋，外立面的多孔结构和几何造型将保护建筑不受冷风的侵袭。这些作用最终产生了微气候，使建筑成为了城市气候的调节器（图7~图10）。

植被"爬上"建筑，与规划功能交织形成了一座垂直式花园。外立面与气候条件相互影响，汇聚了太阳能且重新组织了气流；而外立

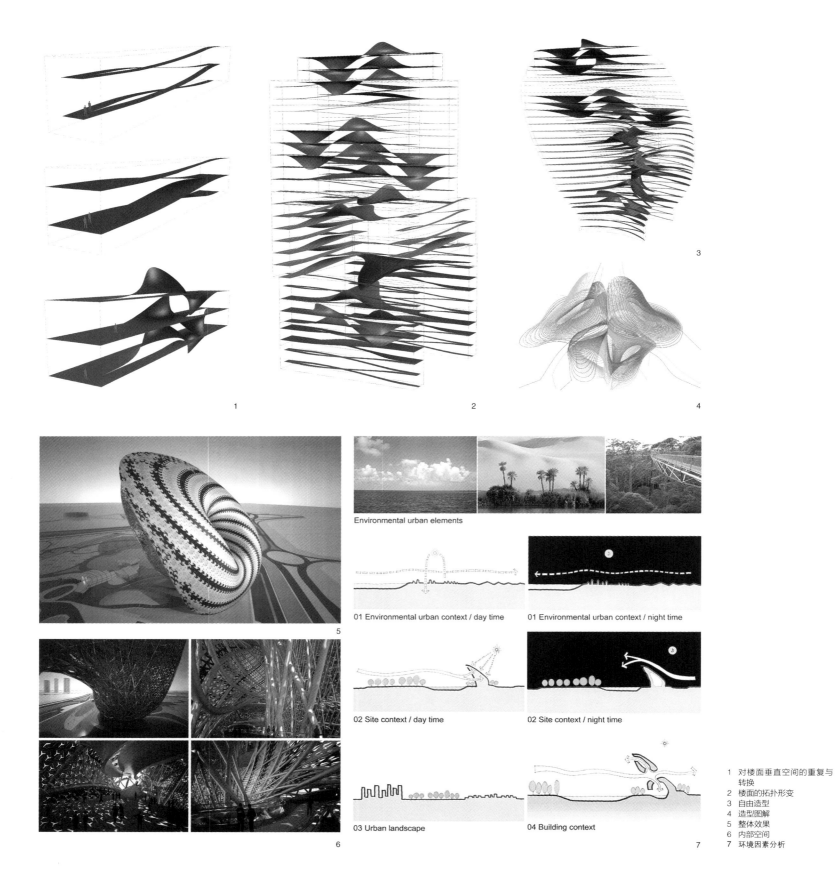

1 对楼面垂直空间的重复与转换
2 楼面的拓扑形变
3 自由造型
4 造型图解
5 整体效果
6 内部空间
7 环境因素分析

高层建筑 | 理论研究

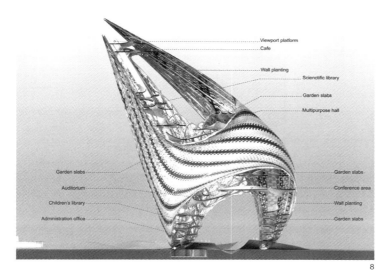

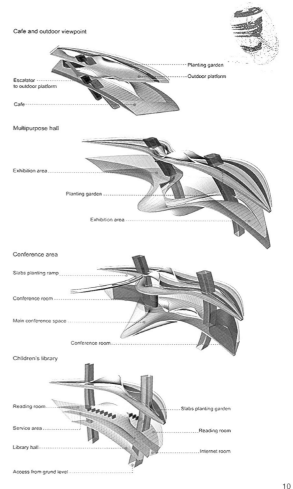

8 立面
9 与环境融合的建筑
10 结构体系

面的内侧则与花园协调共生。毫不夸张地说，楼面和墙体的种植系统孕育了一座花园。外立面矫健挺拔、棱角分明，承担着适应天气变化的功能；而内侧绿草成荫、生机盎然，资料库、会议中心、礼堂、咖啡厅、展览区和花园等各项功能分置其中（图11～图13）。

该工程最为显著的特点就是整座建筑中人与自然的共处，这体现了设计者们对建筑承载自然以及彼此相互影响的深刻理解。

五、可持续性

可持续性是当今建筑设计领域主要关注的议题，而以上探讨的所有问题无一不是在解决可持续性问题——环境、经济以及社会的可持续性发展。因此，不仅要设计出高效节能的建筑，而且要融入可持续性设计理念，令大众了解和体会到建筑与环境的相互作用和影响。超高层建筑正因其巨大的建筑体量，才拥有了这伟大的契机——展现可持续性发展的优势和魅力，进而成为社会发展的标志。

六、"少即是多"的指导思想

我们多次重申"少即是多"的指导思想，需要明确两个重要目标。首先是要在建筑领域内营造一个动态协同合作体系，这样知识和技术就能够在大型建筑事务所、工作室以及学术性机构之间进行快速的共享和流动。我们作为建筑师这一群体，势必要紧跟潮流，融入其中，否则就将被淘汰。其次，也是更为关键的一点，便是要将建筑这一科学领域向其他学科开放。建筑必须同其他学科展开互动与交流，有意识地了解其他学科领域的研究方向和最新成果，这样才能将建筑业推向一个崭新的阶段。建筑咨询的时代已经成为过去，协同合作的时代正在到来。

作者简介

山扬·李沧 Grimshaw 建筑事务所主持设计师

安德烈斯·阿利亚斯·马德里 扎哈·哈迪德建筑师事务所项目设计师

何塞·拉蒙·特拉莫耶雷斯 GGLab 主持设计师

HIGH-RISE BUILDING | THEORETICAL RESEARCH

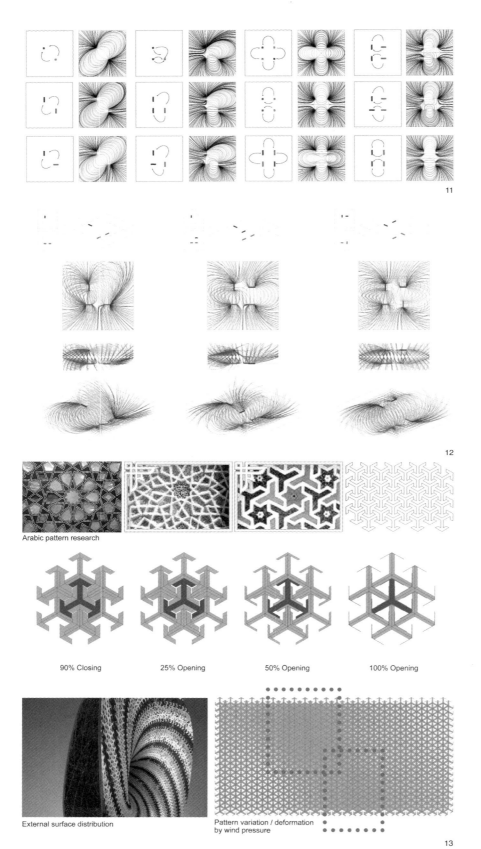

Arabic pattern research

90% Closing   25% Opening   50% Opening   100% Opening

External surface distribution

Pattern variation / deformation by wind pressure

11 形态拓扑形变图解1
12 形态拓扑形变图解2
13 从伊斯兰符号演变的结构立面

THINKING OF THE SUPER HIGH-RISE BUILDINGS FROM THE URBAN PLANNING VIEW
基于规划视角的超高层建筑思考

吴婷婷 王世福 邓昭华 | Wu Tingting Wang Shifu Deng Zhaohua

国家社会科学基金重大项目（11&ZD154）
国家自然科学基金项目（51208202）

目前全球最高的100座超高层建筑中有六成在中国，超高层建筑已经形成一股强劲的潮流，起此彼伏的高度竞赛不禁让人思考"高度"到底意味着什么？本文试图从建筑本身的物理高度相关因素出发，全面理性地认识建筑高度的内涵，建立更客观的超高层建筑价值判断，以期城市规划对超高层建筑实施更有效地控制与引导。

一、高度的经济属性——关于超高层建筑的高度适宜性

随着高度的提高建筑造价上升是一般规律，其提高的程度是存在拐点的。超高层建筑经济造价影响因子主要包括地下基础、结构选型、机电选型、电梯数量、幕墙形式等，其中最关键的是结构选型，按一般建造经验，200~250 m和450~500 m的高度是两道门槛，跨越门槛需要选择不同的结构方式，一般会形成造价的阶梯式抬升。从经济角度来看，因超高层建筑在设计和材料使用上的要求苛刻，尤其是对消防、防震、防风等指标要求很高，因此造价极其昂贵，通常高度200 m以上的高楼每平方米建安成本（不计税）基本要在1万元人民币以上，并且后期使用还要在能源消耗、垂直运输、管理运营和维护等方面持续投入高额成本。

假设在相同的城市区位，建筑的高度越高代表可出让面积越大、收益越高，其形象的凸显度也在一定程度上影响着租金水平。从广州珠江新城核心商务区22个超高层建筑样本中可见，建造规格和建设年限相近，高度因素确实与租金收益正相关，高度的价值认可是实际存在的（图1）。但随着高层建筑在高度上的提高，建筑内部结构面积比例也随之大幅增加，使得内部空间的实用率缩小，因此，总收益水平呈上升趋缓态势。

从经济属性的分析模型可以推断，超高层建筑的建造存在价值拐点，一旦超越某个高度，就会出现收益空间大幅缩窄甚至倒挂的情况（图2）。此外，关于超高层建筑的使用舒适性随着高度的上升而下降也是普遍共识的问题。因此，理性看待高度适宜性应该从多个角度进行分析，尤其关注经济的价值拐点。超高层的建设一旦失去经济意义也就失去了节地的意义。同时，在规划控制中关于建筑高度的设定也不能简单从形态美学角度出发，宜适当考虑经济和社会因素，增加市场可行性，避免实施过程中不必要的调整。

二、高度的权利属性——关于高度的法理思考

建筑体积的形成涉及容积率和高度这两个重要指标。容积率指向建筑容量，是衡量开发利益的核心指标，高度则是表达建筑形态的关键。以一个CBD典型的超高层商务地块为例，1 hm²用地，控制在容积率10，按照一般规律标准层1 500~2 000 m²（超过2 000 m²需要增加防火分区），假设高度不受约束，那么可建设的高度区间为300~450 m，其中有100多米的弹性幅度。就目前的城市实践而言，超高层建筑的空间资源尚不可言之稀缺，但建筑高度差距的背后实际存在着巨大的财产属性差异。

如果不考虑经济因素，这部分高空空间涉及到的外部性和公共性的差异是敏感而不可忽略的。超高层建筑体量巨大、人口集中，给城市交通、水电、消防等形成巨大压力，人员承载量增加还使得地区内开敞空间的人均得地率下降，其外部性显而易见。在通风、采光、视线、微气候、交通、配套设施、光污染等方面也对相邻地块造成环境的不公平，如果相邻地块是居住用途则矛盾更为尖锐。

随着人地关系的日益紧张以及城市的立体化发展，空间利用权逐渐受到各国关注。《中华人民共和国物权法》（2007）明确了空间使用权，其中第一百三十六条规定"建设用地使用权可以在土地的地表、地上或者地下分别设立。新设立的建设用地使用权，不得损害已设立的用益物权"，也就是说，法律规定了一定范围的立体空间可依法占有、使用和收益，具有经济价值与排他性。但是，关于空间权是否可以与土地所有权和使用权分离而成为单独的用益物权，在法学界尚未形成统一的认识。[1]

台湾学者温丰文认为空间权可以分为空间所有权和空间利用权，而空间利用权又可以分为空间地上权、空间役权、空间租赁权、空间使用借贷权，即空间权可以与土地分离而独立存在。[2]这一思维的转变与当今快速发展的空间利用与建造技术不无关系，高空或地下穿越的建筑物寻常可见，典型的如地上地下的轨道交通，日本还出现高速公路直接穿越高层建筑的案例，反映了空间利用日益复合化的趋势。因此，可以推断，空间权与开发权结合可对同一标的物进行三维分割将是未来空间权研究的重要方向。建筑高度问题不但涉及空间权也关系到周边用地物权，其清晰的界定、合法的调整是城市规划的当然职责。

中国目前对于空间权的认知在规划控制和财产界定中均存在模糊点。例如，层数控制和顶部高度控制在各地的规划控制中并存，本质上的差别在于对权利的指向不同：顶部高度控制反映了建筑物对高度空间的占有，是对空间权的明确，而层数控制则由于层高的多种可能性，对空间权的界定较为含糊。又如，过去由于缺乏对建筑层高的明确规定，出现了许多"小复式""N+1"等产品的偷增面积现象。2011年出台的《广州市规划管理建筑面积计算办法》规定，住宅建筑层高大于3.6 m且小于或者等于5.8 m的，按照该层水平投影面积的2倍计算建筑面积，大于5.8 m的按3倍计算建筑面积。各地城市也陆续出台相关规定，如北京市大于4.9 m、柳州市大于4.5 m、吉林省大于4.8 m算2倍面积。这表明了对空间权经济属性的强化认识过程，但是关于高度的财产属性仍然被忽视，简单的例子如房产证上关于层高信息的缺失，2.8 m层高的住宅的舒适度显然低于3.3 m层高，房产价值的差异并未在房产证上得以显示。

三、高度的公共属性——天际线的公共价值

目前的城市规划实践中，因满足公共利益需要而行使的绝对高度控制，包括基于飞机飞行的限高管制、历史城区保护、临近自然的周边需要等情况，在一般城市地区，建筑高度经常成为讨价还价的焦点。建筑高度，尤其是超高层建筑高度对天际线的形成影响重大，构成了城市特色意象，蕴含了整体景观价值，具有重要的公共属性特征。

天际线的价值维育既包括保护也包括形塑，但是什么样的天际线是好的不易界定，各个地方对审美的理解存在差异。如对城市天际线与山体构成的图底关系，各地的审美取向就存在显著差异，如顺应山势、强化山势、与山势反向等。但是一般来说，观赏度较强的天际线构成均满足静态美学的几个特点，如对比、韵律、和谐、层次等。城市地标可能是低矮伸展的公共建筑，也可能是冲破云霄的超高层建筑，对于天际线的控制关键在于主从关系及整体趋势。

天际线控制一般有两种常规手段，一是垂直视角，主要是针对历史保护区及周边的高度控制，另一种是人体水平视角，体现从主要开敞空间眺望目标地区的视野。在2013年全球最美城市天际线排名(The World's Best Skylines)中①，前12名城市中心区绝大多数都享有滨水或大绿地的开敞面，并且与超高层地标建筑紧密相邻（图3）。可见，开敞面对于高层密集中心区形象的感知具有重要意义，地标建筑的空间位置关系是天际线形塑的关键。

广州市CBD珠江新城的地标建筑东塔、西塔就曾经历了位置的重大调整，从远离滨水地区的城市主干道南移至滨江一线的公共文化地标地段。另外，正在规划和建设中的广州国际金融城也出现了相似的决策调整，起步区两个300 m的超高层地标地块远离珠江，从滨江水平视角天际线来看，300 m高楼与前侧方200 m高楼的视觉高度几乎一致，业主单位拿地后均提出了维持开发总量但降低建筑高度的要求（图4）。事件所反映的心理是，与其不被认为地标，不如降低高度的经济成本。两个案例反映的共性问题是，超高层地标建筑是经济成本与公共属性的博弈结果，与大尺度开敞空间应构成积极的空间关系，这类天际线的公共属性问题是规划设计过程应当着重考虑的因素。

此外，不得不提的是，商务商业与住宅的超高层建筑在公共属性上是有巨大差别的，产权私有的超高层住宅无论多高，都不是代表城市形象的"地标"，广州珠江滨水地段滨江东路的屏风式高层住宅即是典型一例。对于水面、山体、广场绿地等重要的城市公共开敞空间，超高层住宅应当适当后退并形成足够的退让坡面，以及适当控制面宽和必要的视线通廊。从伦理上分析，规划认为这部分退让体积是对公共开敞空间品质的保护，具有维护公共利益的性质。

四、地标建筑的城市意义

一个城市的超高层建筑建造动力主要来源于两方面：从建造者角度，显赫的地标建筑展示了企业形象、实力和信心，同时带来了物业收益的诱惑；从城市角度，超高层提高了城市的知名度，是全球化背景下城市营销的重要载体。近年来，绿地集团把20多栋超高层建筑遍插大江南北，尤其青睐于二、三线城市，这些建筑大都成为城市地标，为城市建造地标换取土地的开发模式是该企业的发展策略，反映了"做政府想做的事情"的思维逻辑，值得思考。积极地理解，超高层建筑隐含城市繁荣、国际化的象征，同时也是反映城市内源动力膨胀的一种政治经济现象，可认为城市通过自身空间资本的积累争取对外在发展动力的吸引；消极地认识，炫耀式的形象可能是一种假象，建筑高度与经济水平之间存在不协调的堕距。

客观而言，成为一个城市地标的超高层建筑必然反映了技术的先进及经济的实力，提升了城市的知名度和识别性。超高层建筑数量、规模和经济发展速度正相关，表达了城市对经济发展的热切愿望。通过超高层地标与城市经济发展水平GDP、城市竞争力等要素的关系梳理可以看到，目前中国的超高层建筑拥有与城市实际经济水平存在一定程度的偏差（图5）。

回望历史，从巴比塔到芝加哥第一座超高层建筑，表现了人类征服自然发展和科学技术的雄心壮志。19世纪末美国率先引领了超高层建筑的建造浪潮，到21世纪初亚洲的摩天大楼热，此起彼伏的高度竞赛直白地反映了城市强烈的竞争意识。这些超高层地标建筑肩负了推销城市的重任，尤其是快速发展中的城市在超高层建造方面显得不遗余力，从全球超高层建设总量前50名的城市分布来看，亚洲城市占了一半以上。

欧洲城市的超高层建造数量及规模都不大，远低于亚洲和美洲。在伦敦的金融城，超过200 m的超高层仅有两栋，巴黎拉德芳斯至今未有超过200 m的超高层，并且建设都远离老城区。表面上看是由于历史保护和传统城市形象在欧洲的认可度很高，但深层原因是发展阶段的差异，欧洲工业化与城市化的进度早于亚洲，现阶段的内在发展动力亦低于美国，也没有像中国因供需关系突出而产生膨化效应，因此城市发展平稳，形态演进缓慢。超高层建筑的城市景观现象的形成是一个长期的、不断变化的、由多个系统和不同时期相互交织的复杂过程，是一种反映了抽象的政治经济过程的结果。

五、超高层地标建筑的公共政策

前文论述了天际线的公共价值以及地标建筑的城市意义，说明了超高层建筑尤其是地标建筑无论在城市景观表现还是城市内涵表达上都具有一定的公共属性。[3]因此其功能设计应该适当地考虑开放性。

高层建筑 | 理论研究

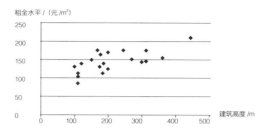
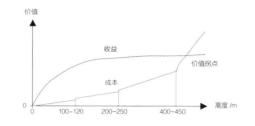
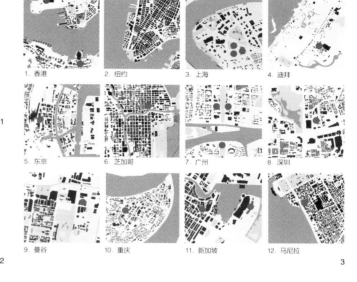
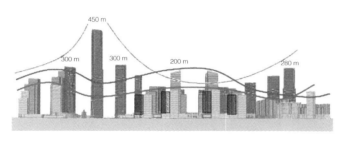
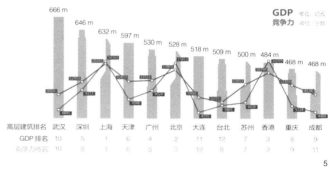

1 珠江新城核心商务区建筑高度与租金关系（图片来源：作者自绘）
2 超高层建筑建设成本与收益模型（图片来源：作者自绘）
3 地标建筑与开敞空间位置关系（图片来源：作者自绘）
4 广州国际金融城天际线（图片来源：《广州国际金融城起步区控制性详细规划》）
5 超高层地标与经济发展水平GDP、城市竞争力的关系（图片来源：参考2013中国城市综合竞争力排行榜、2013年中国城市GDP排名、2013中国高楼排行榜绘制）

另外在城市整体空间形态框架上，超高层建筑的显现度很高，其视野是一般建筑无法企及的，可以认为是对高空景观资源的占有，并且越高的建筑占有率越高，因而适当返还公共利益是合乎法理的。在规划管制上，宜鼓励其在地面层提供更多的积极城市空间，将顶层向公众开放，在许多欧美国家都有类似的政策引导。

另外，基于地标建筑的公共性与影响力，其规划与设计尤其应当进行公众参与。目前中国的公众参与的水平有限，处于"梯子理论"的较低阶段，但是随着法规制度及保障措施的不断完善，以城市居民为主体的力量将逐渐成为城市发展决策的重要影响之一。建筑设计的公开评审制度将逐步完善，由当前单一的精英模式向精英与公共参与结合的模式转变。在珠江新城的案例中，早期的规划方案规定东塔与西塔高度基本一致，但由于建设阶段不同，西塔于2010年落成，高度为440 m，东塔后于其建设，当时政府又提出不限高、不限容积率的招标方式，导致高度问题经历多轮迂回的精英决策，从高于西塔10%，到高于50 m、100 m。设想如果公众参与介入，以当下网络社会开放和自由的主流思想，对于城市地标的高度问题必然有超出精英决策的巨大包容，其结果可能更富于想象力和多元化，同时也可满足公众知情和参与的需求，是值得尝试和探索的。

注释

① 详见 http://www.skyscrapercity.com。

参考文献

[1] 陈祥健. 关于空间权的性质与立法体例的探讨 [J]. 中国法学，2002（5）：102-108.

[2] 温丰文. 空间权之法理 [J]. 法令月刊，1988（3）：8.

[3] 王世福，周可斌. 浅议城市高层建筑公共性 [J]. 城市建筑，2009（10）：24-26.

作者简介

吴婷婷　华南理工大学建筑学院博士研究生

王世福　华南理工大学建筑学院教授，博士生导师
　　　　华南理工大学亚热带建筑科学国家重点实验室

邓昭华　华南理工大学建筑学院讲师
　　　　华南理工大学亚热带建筑科学国家重点实验室

FIELD RESEARCH ON THE SPACE FORM OF SHENZHEN'S HIGH-RISE BUILDINGS
深圳高层建筑空间造型实态调研

覃力 刘原 | Qin Li Liu Yuan

深圳是个非常年轻的城市，同时，也是一个典型的高密度城市，其经济发展速度和城市建设速度都令人惊叹。深圳的GDP总量长期位于中国的前列，人均GDP排名第一，据权威机构测算，到2015年，深圳的GDP将会继上海、北京之后超过香港。深圳的居住人口已经达到1 300多万，但城市面积却相对狭小，仅有1 953 km²，是上海的1/3，广州的1/6，北京的1/8，可以说是中国密度最高的一座城市。这样的经济发展速度和超高的城市建设密度，必然催生出大量的高层建筑，所以深圳的高层建筑一直以来都受到各界的关注（图1）。本文尝试从发展进程、空间、造型等几个方面，对深圳高层建筑的现状进行初步的分析和探讨。

一、发展进程

1980年以前，深圳还是个小渔村，城市建设十分落后，建筑物大都是平房，最高的只有5层。深圳市成立之后，经济与城市建设极速增长，高层建筑迅速发展起来。1982年竣工的第一座高层建筑电子大厦（图2）高20层，是当时深圳的第一高楼。继电子大厦之后仅仅3年，深圳国际贸易中心大厦落成，大厦高160 m，53层，是当时中国最高的建筑，同时，该建筑的施工采用独创的"滑模"技术，创造了3天一层楼的"深圳速度"，首次让深圳的高层建筑受到了全国上下的瞩目。深圳国际贸易中心大厦的落成，标志着深圳的高层建筑在较短的时间内便达到了较高的水平，并且开始了高度上的引领。

随后，一批高度超过100 m的高层建筑陆续建成，据统计，至1985年夏，深圳市批准兴建的高层建筑已达297幢。从1981年至1987年的短短7年时间里，深圳市建成18层以上的高层建筑140幢，其中超高层建筑66幢，可见深圳高层建筑发展势头的迅猛。

这一时期，由于罗湖区拥有火车站及靠近香港的地理优势，因此建设速度特别快，受香港高层、高密度建设理念的影响，只用了几年时间便形成了当时深圳高层建筑最集中、最密集的"罗湖商业区"，那一时期深圳的大多数高层建筑都集中在这一区域，"罗湖高层建筑群"也成了那个时代的象征。

到了20世纪90年代，邓小平的"南方谈话"肯定了深圳在过去10年里取得的成就，并提出"加大改革力度""改革开放的胆子要大一些"的要求，进一步促进了深圳的经济发展。加之其时境外设计机构及国际知名建筑师也开始关注并参与到深圳高层建筑的设计建设之中，这一时期深圳高层建筑的建设可说是高潮迭起，曾一度代表着中国高层建筑设计、施工等方面的最高水平。

1992年建成的深圳发展中心大厦（图3），是中国第一栋大型高层钢结构工程，将钢结构引入超高层建筑，在当时的中国建筑界也产生了很大的影响，印有深圳发展中心大厦照片的年历在中国各地都能够见到。而最具影响力的要数1996年建成、高384 m的地王大厦（图4），是当时中国的第一高楼，直至2011年10月，一直保持着深圳最高楼宇的纪录。地王大厦的施工以两天半一层的建设速度，刷新了之前深圳国际贸易中心大厦创下的3天一层楼的"深圳速度"，其高宽比也创造了当时世界超高层建筑最"扁"、最"瘦"的纪录，由此成为20世纪90年代深圳高层建筑的代表，并将中国的高层建筑设计和建设推向了国际水平。

从整体上看，深圳高层建筑的建设在此期间有了进一步的发展。据不完全统计，到1996年6月，深圳市范围内竣工和在建的18层以上的高层建筑就已经达到了753幢。到20世纪90年代末，深圳建成的高度100 m以上的非住宅类超高层建筑24幢，高度150 m以上的摩天大楼17幢。

20世纪90年代后期，深圳的高层建筑又展现出与高科技、智能化相结合的发展趋势，如深圳发展银行大厦（图5）、特区报业大厦等。此后，深圳的高层建筑设计更加注重新技术、新理念的应用，在技术上也取得了一些突破，某些方面处于世界先进行列。例如2000年建成的赛格广场（高292.6 m）（图6），是由中国自行设计和总承包施工的高智能超高层建筑，同时，也是目前为止世界上最高的钢管混凝土结构大楼。

21世纪以后，福田中心区的建设日趋成熟，在城市设计和法定图则的指引下，落成了大批很有特色的高层建筑，如深圳电视中心（高121.9 m）（图7）、深圳国际商会中心（高218 m）、中国联通大厦（高100 m）（图8）、安联大厦（高159.8 m）、新世纪中心（高195 m）、卓越时代广场二期（高218 m）、诺德中心（高198 m）、中国凤凰大厦（高109 m）、深圳卓越世纪中心（高280 m）（图9）、深圳证券交易所新总部大楼（高245.8 m）等。目前，福田CBD超高层建筑群已经基本形成，并与罗湖区一同成为深圳市高层建筑最为集中的区

1 深圳全景（图片来源：http://www.gaoloumi.com）
2 电子大厦（图片来源：作者自摄）
3 深圳发展中心大厦（图片来源：作者自摄）
4 地王大厦（图片来源：http://image.baidu.com）
5 深圳发展银行大厦（图片来源：作者自摄）
6 赛格广场（图片来源：http://image.baidu.com）
7 深圳电视中心（图片来源：作者自摄）

域，同时，也成为深圳迈向现代化国际大都市的外在表征。

2011年建成的深圳京基100（图10），以441.8 m的高度，打破了地王大厦保持了15年的高度纪录，成为深圳第一高楼。京基100当时在全球高楼排行榜中排名第九，再次使深圳在新一轮的高层建筑建设热潮中走在了前面，为深圳带来了全新的时尚气质与领先理念。正在施工建造中的深圳平安国际金融中心的高度，更是达到了660 m，建成之后将超过上海中心，成为中国第一高楼。

进入21世纪以来，深圳的高层建筑在数量和质量上都呈现出质的飞跃。高层建筑在城市空间的分布上更加集中，在建筑与城市的关系上，也更加注重通过城市设计来加强高层建筑与城市空间之间的互动关系，从群体上调控建筑立面、高度、色彩、材质等要素。特别是2010年以后，随着卓越世纪中心、京基100、大中华国际金融中心、中洲中心、京基滨河时代广场等高层建筑城市综合体的建设，深圳的高层建筑越来越趋向于集群化设计建造的组群关系，向着建筑城市"一体化"的方向发展。而随着新时期城市规划的调整，深圳又在策划前海、后海以及红树林等大规模超高层建筑集中区域，红树林湾区将形成拥有众多三四百米高建筑的"超级城市"景象。

与此同时，"关外"的宝安、龙岗也在规划建设大量高层、超高层建筑，突破了原来只有"关内"才能见到超高层建筑的情况。已经建成的有宝安区的荣超滨海大厦、国际西岸商务大厦、翰林大厦，龙岗区的正中时代广场、珠江广场、银信中心（图11）等，在建中高度超过250 m的还有中粮大悦城、壹方中心、春天大厦和深圳佳兆业城市广场等，可见"关外"的高层建筑也在迅速发展。

据《摩天城市报告》数据显示，到2014年，深圳已建成高度200 m以上的超高层建筑有39座，数量仅次于上海，位居中国第二，在未来的5年内，高度200 m以上的超高层建筑将达到102座，超过上海和香港。届时，高度300 m以上的摩天大楼会有30座，深圳现在的第一高楼京基100，在高度排行榜中也只能名列第四。排在前三名的是666 m高的蔡屋围晶都片区改造项目、660 m高的深圳平安国际金融中心和518 m高的佳兆业环球金融中心。由此可见，深圳在未来的几年内，还会加快高层建筑建设的步伐，向着高度更高的方向发展。

二、空间构成的尝试

高层建筑设计得如何，并不单纯取决于建筑高度。空间效果更为关键。尽管在技术方面，高层建筑的空间组织方式会受到很多限制，而且随建筑高度的增加，限制也会越来越苛刻。但是，经过30多年的发展，深圳的高层建筑在空间构成上仍做出了不少有益的尝试，有着一定的示范性作用。

在高层建筑的空间构成上，垂直交通体——核心筒起着非常重要的作用，核心筒的布置方式决定着整体的空间效果。最简便、经济的空间构成方式是中央核心筒式的布置形式，不过，受国际先进理念的影响，高层建筑的设计中也开始强调人性化和空间的多样性。

因此，20世纪90年代之后深圳的高层建筑中，各种不同的空间构成方式不断地涌现，从单核心筒发展到双核心筒、分散核心筒以及核与主体分离的形式，从中央核心筒发展到偏置核心筒。例如特区报业大厦、汉唐大厦等，即采用了双侧核心筒的空间构成方式；本元大厦、中国联通大厦、深圳发展中心大厦、汉京国际大厦等，采用了单侧外核心筒的空间构成方式；银信中心、方大大厦，则采用了分散核心筒的空间构成方式；而在建的由汤姆·梅恩（Thom Mayne）设计的350 m高的汉京中心，更选用了核心筒与主体分离的方式，可谓超高层建筑设计中的一大创举（表1）。

表1 高层建筑核心筒布置方式

	概念示意	标准层平面实例	
中央核心筒式		现代国际大厦	地王大厦
双侧外筒式		汉唐大厦	特区报业大厦
单侧外筒式		汉京国际大厦	本元大厦
分散核心筒式		方大大厦	银信中心
分离核心筒式		汉京中心	

空间构成方式上的这些变化，不仅带来了建筑造型上的改变，还形成了许多新颖的空间效果。相对于中央核心筒的布置方式，双侧核心筒的空间构成方式可以在高层建筑的底部形成开敞通透的大空间。而核心筒分散到外侧，不仅改善了高层建筑内部公共空间封闭昏暗的弊端，使电梯厅和走廊能够直接对外采光通风，而且还可将电梯设计为景观电梯，使人流动线可视化，给人以完全不同的感受。

除了核心筒的布置方式之外，深圳的许多高层建筑还在追求使用空间内部的变化，在局部设置几层通高的活动空间。比较早的实例有1998年建成的特区报业大厦，该建筑在东南角上设计了一系列3层通高的活动空间，打破了按楼层平均分割、各层互不连通的做法，在局部形成空间上的变化。近些年，这种在高层建筑中植入异质空间的方法已经得到了广泛应用，深圳许多高层办公楼内部都有所谓的复式空间，一些公司还将几层通高的大堂搬到了空中，形成空中大堂，增加垂直方向上空间的变化，例如闽泰大厦中的华森公司的大堂、卓越大厦中的卓越集团的大堂，以及中国联通大厦上部高达12层的开放式中庭。刚刚建成的深圳证券交易所新总部大楼，更是将交易大厅等公共活动空间搬到了30多米高的空中（图12）。

有些高层综合楼的酒店也流行空中大堂的做法，为了争取更好的景观效果，将几层高的大堂和餐厅布置在塔楼的上部，可以居高临下地欣赏城市景观。例如华润中心的君悦酒店、京基100的瑞吉酒店，都是在顶层寻求空间变化的实例。瑞吉酒店的大堂设置于京基100的94层，是一个由玻璃覆盖的高达40 m的空中大厅，大厅的上部设计了一个鹅蛋形的餐厅，夜晚灯光开启之后，晶莹的大厅与城市夜景互为观赏对象（图13）。

在高层建筑中设置空中花园，是一种借助生态理念、富有地域特色的做法，还可以通过竖向空间的变化改善外观效果，在深圳的高层建筑中也运用得比较多，典型案例有安联大厦（图14）、本元大厦、汉京国际大厦、海岸城东座写字楼等。在建筑的不同方向、楼层设置空中花园，不仅可以种植绿化，将自然景观引入到高层建筑中，为人们提供空中的室外化空间，改善高层建筑内部的微气候，同时，又可增添休憩交流的空间，打破建筑外观的刻板形象，避免高层塔楼千篇一律的单调感。

安联大厦每隔4层，在建筑的南、北两侧各设置了6个空中花园，在东、西两侧各设置了8个空中花园，把阳光、新鲜空气和绿色植物引入了超高层大楼之中，形成建筑中的"绿肺"；本元大厦沿着建筑垂直方向，一共设置了28个空中花园；田厦国际中心则是在建筑的东、西两端设置了两个3层高的空中花园。

还有一些高层建筑的空中花园的设置是随着楼层的更替而产生变化的，形成一定的韵律，使空间既丰富又协调统一（表2）。汉京国际大厦中虽然空中花园都在建筑的同一方向，但按奇偶层交替变换位置之后，形成了一种灵活丰富的空间效果。空中花园的使用率很高，人们在无法接触到地面的空中，仍然可以体验到庭园的感受，工作中发生的一些间歇性行为（如打电话、接待客人、讨论问题等）也可以在空中花园进行。

表2 空中花园在高层建筑中的位置关系

项目名称	田厦国际中心	安联大厦	汉京国际大厦
标准层平面			

8 中国联通大厦（图片来源：张一莉的《深圳勘察设计25年：建筑设计篇》一书）
9 卓越世纪中心（图片来源：http://image.baidu.com）
10 京基100（图片来源：http://image.baidu.com）
11 银信中心（图片来源：作者自摄）
12 深圳证券交易所新总部大楼（图片来源：作者自摄）
13 京基100上部中庭组合图（图片来源：TFP事务所《深圳京基100》一文）
14 安联大厦空中花园（图片来源：作者自摄）
15 汉京中心（图片来源：http://image.baidu.com）

三、形式特征的演变

深圳的高层建筑都是在20世纪80年代以后建成的。80年代中国刚刚改革开放，受现代主义建筑思潮的影响，深圳这一时期建筑的主要特点与中国其他地区一样，多是在建筑形体上追求体积感和几何关系，以均质的外墙、铝合金窗以及玻璃幕墙作为表情符号。在经济条件有限、"形式追随功能"的时代，高层建筑的形式更多是服从功能的需要，对功能和结构进行真实表达。因此，"盒子形"的高层建筑在当时较为流行，如1982年落成的电子大厦、分别于1985年和1986年建成的深圳国际贸易中心大厦与深圳国际金融大厦、1989年建成的华联大厦等均采用"盒子形"建筑形式。

深圳国际金融大厦采用现代的建筑风格，四个角部设置了菱形柱子，巨大的体积和大面积的深蓝色玻璃幕墙形成鲜明的虚实对比，其出色的整体造型、合理的空间布局和明确的功能分区，在时隔28年后的今天，仍旧风采依然。在此期间，深圳的高层建筑设计主要跟随中国的大趋势，加之深圳高层建筑的设计师与建筑设计机构多来自于中国各地，因此，建筑形式带有天南地北的地域特点，呈现出多样化的高层建筑形式表达，并未形成深圳本土的特色。

20世纪90年代深圳的高层建筑已开始从高层向超高层方向发展。随着经济实力的增强和技术水平的提高，人们渐渐厌倦了"国际式风格"的盒子形造型，建筑师们开始追求个性化的创作，"标志性"成为高层建筑设计的精神诉求。这段时间建成的高层建筑，表情比较丰富，个性张扬，造型感较强。如海王大厦尝试极具穿透力的架空处理方式，以及在立面上装饰雕像的做法；深圳发展银行大厦根据基地所处的位置，将建筑设计成由西向东逐步升高的阶梯状，巨大倾斜向上的构架带有"高技术"的审美趣味，寓意"发展向上"，是当时深圳最具特色的建筑之一；佳宁娜友谊广场、深业大厦和联合广场等，也都是这一时期高层建筑的典型代表。

这一时期的高层建筑设计常在屋顶上做文章，以凸显其标志性，例如世界金融中心就在塔楼的顶部设置了层层叠加的尖顶，招商银行大厦设计了倒锥体的屋顶，特区报业大厦的屋顶装饰更是高达50多米。但是深圳的高层建筑不论屋顶如何处理，却较少采用传统的坡屋顶形式，只有蛇口的海景广场等一两幢建筑应用了四坡屋顶。

与20世纪80年代高层建筑基本上由中国内地建筑师设计不同，从90年代开始，香港及国外建筑师也参与了一些重要的高层建筑设计。

如深圳发展中心大厦由美国锡霖集团和香港迪奥设计顾问有限公司设计，地王大厦由美籍华人张国言设计，招商银行大厦由美籍华人李明仪设计。境外设计公司的引进，带来了一股新风，增加了国际间的交流，使高层建筑的设计水平有了很大的提高，也使得深圳的高层建筑设计更加趋向于国际化。

进入21世纪之后，由于深圳作为新城市的窗口作用和开放性，促进了国际间的交流合作，使得新理念和新技术不断地涌入，本土建筑师的素养也在实践锻炼中得到了提高，设计水平进步很快。在此期间，由于更加重视从城市的总体空间效果来把握高层建筑的设计，所以，高层建筑的体量趋于规整，形式趋于纯净，不再追求张扬的个性，仿佛又回到了方盒子时期。与之前不同的是，建筑立面不再强调虚实对比，色彩也比较素雅，呈现出一种注重整体性和肌理感的"表皮"特征。许多高层都将设计的侧重点放在了建筑的表层，并结合幕墙表面肌理和标准层空间的局部变化，来达到整体效果的出新，立面表达更加自由。

与北京、上海等地相比，深圳在高层建筑的设计创作中，缺少了历史文化上的积淀，更多地强调了时代特征。但也许正是这种缺失，没有给建筑师带来更多的束缚，反而提供了一个相对宽松的创作环境，使深圳的高层建筑设计摆脱了民族情结，更加务实，更加讲究效益，追求"现代性"。而经过30多年的积淀，深圳也正在形成自己的文化感染力。

这一时期，更多知名的境外设计公司和设计师参与了深圳高层建筑的设计，如SOM建筑设计事务所设计了新世界中心、招商局广场，RTKL国际有限公司设计了华润中心，Larry K Oltmans参与设计了卓越世纪中心，汉京中心由汤姆·梅恩设计（图15），TFP事务所设计了京基100，OMA设计了深圳证券交易所新总部大楼，哈利法塔的设计人艾德里安·史密斯（Adrian Smith）设计了中洲中心，正在建造的深圳未来第一高楼深圳平安国际金融中心由KPF建筑事务所设计。目前相对重要的超高层大楼几乎都由境外设计公司包揽，在高度300 m以上的超高层建筑设计招标中，本土设计单位一般只能与境外公司合作，这种状况说明了建筑市场已经全面开放，未来的竞争还会更加激烈。

当然，大量的高层、超高层建筑还是由深圳本地建筑师设计完成的，本地建筑师也创造了许多优秀的高层建筑，其中具有代表性的有中国凤凰大厦、卓越时代广场一期、中国联通大厦、大中华国际金融中心、腾讯大厦、银信中心、NEO企业大道等。这些建筑，通过简单的几何体、线性组合与精致的细部设计，表现出了一种简约、典雅的建筑风格。

纵观30年的发展变化，深圳的高层建筑设计受折中主义和古典主义的影响相对较小，在长期追求表现时代特征的同时，摆脱了早期现代主义的影响，较少出现炫技的设计手法，也没有不规则曲线和非线性旋转等时尚跟风现象，除了个别建筑因造型过于夸张、采用"土豪金"幕墙而备受诟病之外，大多数的高层建筑设计均趋于理性。深圳的高层建筑，已经从早期迎合改革开放的"试验场"、追求特立独行的阶段，发展至今，更加注重城市空间关系，强调理性和务实的设计表达，并在空间构成、生态理念与环境融合等方面，做出了许多有益的尝试。

参考文献

[1] 深圳市人民政府. 崛起的深圳 [M]. 深圳：海天出版社，2005.

[2] 深圳市人民政府基本建设办公室，深圳市规划局，世界建筑导报社. 深圳建筑 [M]. 深圳：世界建筑导报社，1988.

[3] 深圳市人民政府基本建设办公室. 深圳基本建设之路 [J]. 世界建筑导报，1988（5）：9.

[4] 深圳市建设局，深圳市城建档案馆. 深圳高层建筑实录 [M]. 深圳：海天出版社，2005.

[5] 张一莉. 深圳勘察设计25年：建筑设计篇 [M]. 北京：中国建筑工业出版社，2006.

[6] 深圳市规划与国土资源局. 深圳市中心区城市设计与建筑设计1996-2002[M]. 北京：中国建筑工业出版社，2002.

[7] 深圳市城建档案馆. 深圳名厦 [M]. 深圳：[出版者不详]，1995.

[8] 覃力. 高层建筑空间构成模式研究 [J]. 建筑学报，2001（4）：17-20.

[9] 叶伟华. 深圳新世纪摩天楼设计浅析 [J]. 时代建筑，2005（4）：66-68.

[10] 覃力. 高层建筑集群化发展趋势探析 [J]. 城市建筑，2009（10）：29-31.

[11] 梅洪元，陈剑飞. 新世纪高层建筑发展趋势及其对城市的影响 [J]. 城市建筑，2005（7）：9-11.

作者简介

覃力　深圳大学建筑与城市规划学院教授
　　　深圳大学建筑设计研究院总建筑师

刘原　深圳大学建筑与城市规划学院硕士研究生

设计作品
DESIGN WORKS

052 英国碎片大厦
THE SHARD (LONDON BRIDGE TOWER), UK

058 西班牙费拉Porta Fira双子塔
TORRES PORTA FIRA, SPAIN

062 杭州华联UDG时代广场
HANGZHOU HUALIAN UDG TIME SQUARE

066 巴西圣保罗无限大楼
SÃO PAULO INFINITY TOWER, BRAZIL

072 阿拉伯联合酋长国阿布扎比投资管理局总部大厦
ABU DHABI INVESTMENT AUTHORITY HEADQUARTERS, UAE

078 韩国三星瑞草大厦
SAMSUNG SEOCHO, KOREA

084 英国尖塔
THE PINNACLE, ENGLAND

088 韩国东北亚贸易大厦
NORTHEAST ASIA TRADE TOWER, KOREA

092 韩国松岛国际城住宅
SONGDO INTERNATIONAL CITY, SOUTH KOREA

096 香港理工大学赛马会创新楼
JOCKEY CLUB INNOVATION TOWER, HONG KONG POLYTECHNIC UNIVERSITY

106 阿拉伯联合酋长国The Opus办公大楼
THE OPUS OFFICE BUILDINGS, UAE

110 马来西亚黎明之塔
SUNRISE TOWER, MALAYSIA

116 罗马尼亚多罗班蒂大厦
DOROBANTI TOWER, ROMANIA

122 埃及尼罗塔
NILE TOWER, EGYPT

128 新加坡花拉阁
FARRER COURT, SINGAPORE

134 广州国际金融中心
GUANGZHOU INTERNATIONAL FINANCE CENTER

140 广州东塔
GUANGZHOU EAST TOWER

144 沙特阿拉伯Tadawul证券交易大楼
THE TADAWUL STOCK EXCHANGE TOWER, KINGDOM OF SAUDI ARABIA

150 阿拉伯联合酋长国55°旋转塔
55 DEGREES ROTATING TOWER, UAE

154 深圳证券交易所新总部大楼
SHENZHEN STOCK EXCHANGE HEADQUARTERS

162 美国111 第一大街
111 FIRST STREET, USA

166 法国灯塔高楼
LA TOUR PHARE, FRANCE

170 墨西哥Bicentenario 塔
TORRE BICENTENARIO, MEXICO

176 天津中钢国际广场
TIANJIN SINOSTEEL INTERNATIONAL PLAZA

182 加拿大多伦多梦露大厦
TORONTO ABSOLUTE TOWERS, CANADA

186 意大利商品交易会公司总部大楼
TRADE FAIR CORPORATE HEADQUARTERS, ITALY

192 韩国青罗城市大厦
CHEONGNA CITY TOWER, KOREA

196 丹麦巨拱
CPH ARCH, DK

202 荷兰奈梅亨商务及创新中心 52°
NIJMEGEN BUSINESS AND INNOVATION CENTRE FIFTY TWO DEGREES, THE NETHERLANDS

208 美国赫斯特大厦
HEARST TOWER, USA

214 美国世界贸易中心二号塔
TOWER TWO ON THE SITE OF THE WORLD TRADE CENTRE, USA

220 阿拉伯联合酋长国阿布扎比资本中心塔
ABU DHABI CAPITAL GATE, UNITED ARAB EMIRATES

224 俄罗斯圣彼得堡奥克塔摩天楼
ST. PETERBURGH OKHTA TOWER & CENTRE, RUSSIA

230 土耳其Atasehir Varyap 项目
ATASEHIR VARYAP PROJECT, TURKEY

236 南京河西新城苏宁广场
NANJING SUNING WEST RIVER CITY PLAZA

240 阿拉伯联合酋长国帝国大厦
EMPIRE TOWE, UNITED ARAB EMIRATES

244 阿拉伯联合酋长国迪拜Pentominium 大楼
DUBAI PENTOMINIUM TOWER, UNITED ARAB EMIRATES

THE SHARD (LONDON BRIDGE TOWER), UK
英国碎片大厦

伦佐·皮亚诺建筑工作室 | Renzo Piano Building Workshop

项目名称：碎片大厦
业　　主：Sellar Property Group
建设地点：伦敦泰晤士河南岸
设计单位：Renzo Piano Building Workshop
合作单位：Adamson Associates (Toronto, London)
建筑面积：12.7万 m^2（办公55 277 m^2，餐饮2 608 m^2，酒店17 562 m^2，公寓5 788 m^2，观景1 391 m^2）
建筑层数：72层（办公4～28层，餐饮31～33层，酒店34～52层，公寓53～65层，观景68～72层）
建筑高度：306 m
垂直交通：44部电梯，8部自动扶梯
第一阶段设计团队（规划审批，2000～2003年）：J. Moolhuijzen (partner in charge), N. Mecattaf, W. Matthews with D. Drouin, A. Eris, S. Fowler, H. Lee, J. Rousseau, R. Stampton, M. van der Staay and K. Doerr, M. Gomes, J. Nakagawa, K. Rottova, C. Shortle; O. Aubert, C. Colson, Y. Kyrkos (models)
第一阶段咨询团队：Arup（结构和设备），Lerch, Bates & Associates（垂直交通），Broadway Malyan（顾问建筑师）
第二阶段设计团队：（2004～2012年）J. Moolhuijzen, W. Matthews (partner and associate in charge), B. Akkerhuis, G. Bannatyne, E. Chen, G. Reid with O. Barthe, J. Carter, V. Delfaud, M. Durand, E. Fitzpatrick, S. Joly, G. Longoni, C. Maxwell-Mahon, J. B. Mothes, M. Paré, J. Rousseau, I. Tristrant, A. Vachette, J. Winrow and O. Doule, J. Leroy, L. Petermann; O. Aubert, C. Colson, Y. Kyrkos (models)
第二阶段咨询团队：WSP Cantor Seinuk（结构），Arup（建筑设备），Lerch, Bates & Associates（垂直交通），Davis Langdon（造价），Townshend Architects（景观），Pascall+Watson（车站执行建筑师）
设计时间：2000年
建成时间：2012年
图纸版权：Renzo Piano Building Workshop

　　伦敦桥大厦，也称碎片大厦，位于泰晤士河南岸伦敦桥车站附近，72层的空间里包括多种功能。集火车、公交、地铁为一体的伦敦桥车站是伦敦最繁忙的交通枢纽之一，每天客流量达20万人次。伦敦桥大厦项目正是对伦敦市市长倡导在主要交通节点大力发展高密度建筑的政策的积极回应。

　　大厦的形态由其在伦敦城市天际线的突出位置所决定。与纽约、香港等城市的情况不同，碎片大厦所在的区域并没有林立的高楼。设计参考了停泊在泰晤士河岸船只的桅杆和莫奈的《国会大厦》（House of Parliament）系列画作。

　　建筑纤细的锥体形式适合不同的使用功能，底部较大楼面为办公，中部为公共区域和酒店，上部为公寓，68～72层是最顶端的公共楼层，在距离地面240 m处设置有公共的观景廊，再向上建筑延伸到306 m的高度。多元的使用功能为大厦增添了活力，对于伦敦如此标志性的建筑，其公共可达性尤为重要。

　　8个玻璃"碎片"决定了大厦的外形和视觉效果。整个被动式双层立面采用了超白玻璃，空腔内设置机械卷帘用以遮阳。玻璃"碎片"之间的"缝隙"作为开放通风口为室内提供了自然通风，"缝隙"处的空间在办公层作为会议室或休息室，在公寓层可作为冬季花园，为密封的建筑提供了与外部环境的重要联系。

　　建筑主体结构构件是建筑中心的滑模混凝土核心筒，容纳了主要的设备管道、电梯和疏散楼梯。总共44部单层和双层电梯在地面各入口、车站广场层和主要功能区之间建立了连接。

　　该项目还包括火车站广场和公共汽车站的重建。原有的屋顶将被玻璃穹顶所代替，零售商铺的搬迁使火车站、公共汽车站和出租车候客站之间得以建立视觉连接，两个新建的平面为30 m×30 m的公共广场将成为项目的中心。公共区域的改善对于这片拥挤、被忽略的区域的再生至关重要，并有望促进该地区的进一步重建。

1 从圣保罗教堂看向碎片大厦（摄影：Renzo Piano Building Workshop – ph. William Matthews）

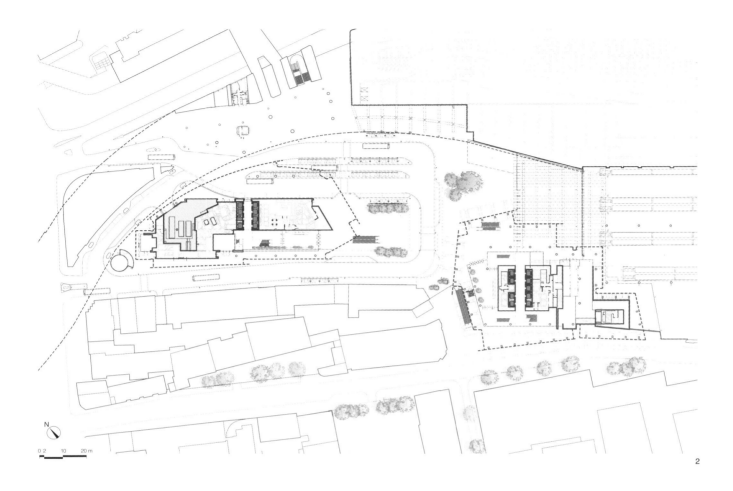

London Bridge Tower, which is also known as the Shard, is a 72 storey mixed use tower located besides London Bridge Station on the south bank of the river Thames. The station, which combines train, bus and underground lines is one of the busiest in London with 200,000 users per day. The project is a response to the Mayors policy of promoting high density development at key transport nodes.

The form of the tower was determined by its prominence on the London skyline. Unlike other cities such as New York or Hong Kong, the Shard is not part of an existing cluster of high rise buildings. References included the masts of ships docked in the nearby Pool of London and Monet's paintings of the Houses of Parliament.

The slender pyramidal form is suited to the variety of uses proposed: large floor plates for offices at the bottom, public areas and a hotel in the middle, apartments at the top. The final public floors, levels 68-72, accommodate a viewing gallery 240m above street level. Above, the shards continue to 306m. The mix of uses add vibrancy to the project: public access was deemed particularly important for such a significant building in London.

Eight glass shards define the shape and visual quality of the tower. The passive double facade uses low-iron glass throughout, with a mechanised roller blind in the cavity providing solar shading. In the "fractures" between the shards opening vents provide natural ventilation to winter gardens. These can be used as meeting rooms or break-out spaces in the offices and winter gardens on the residential floors. They provide a vital link with the external environment often denied in hermetically sealed buildings.

The main structural element is the slip formed concrete core in the centre of the building. It houses the main service risers, lifts and escape stairs. A total of 44 single and double-deck lifts link the key functions with the various entrances at street and station concourse level.

The project also includes the redevelopment of the train station concourse and bus station. The existing roof is to be removed and replaced with a glazed canopy, and retail units relocated to open up visual connections between the train station, bus station and taxi ranks. Two new 30m x 30m public squares will form the centre of the scheme. Such improvements to the public realm are vital to the regeneration of this congested and neglected part of the city and will hopefully provide the catalyst to further redevelopment in the area.

2 车站广场层总平面
3 从圣托马斯街看向碎片大厦
　（摄影：Rob Telford）
4 从圣托马斯街看向碎片大厦
　（摄影：Rob Telford）
5 九层平面
6 二十三层平面

3
4

5
6

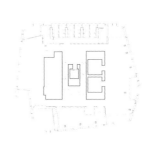

7 南立面
8 三十二层平面
9 三十九层平面
10 六十八层平面

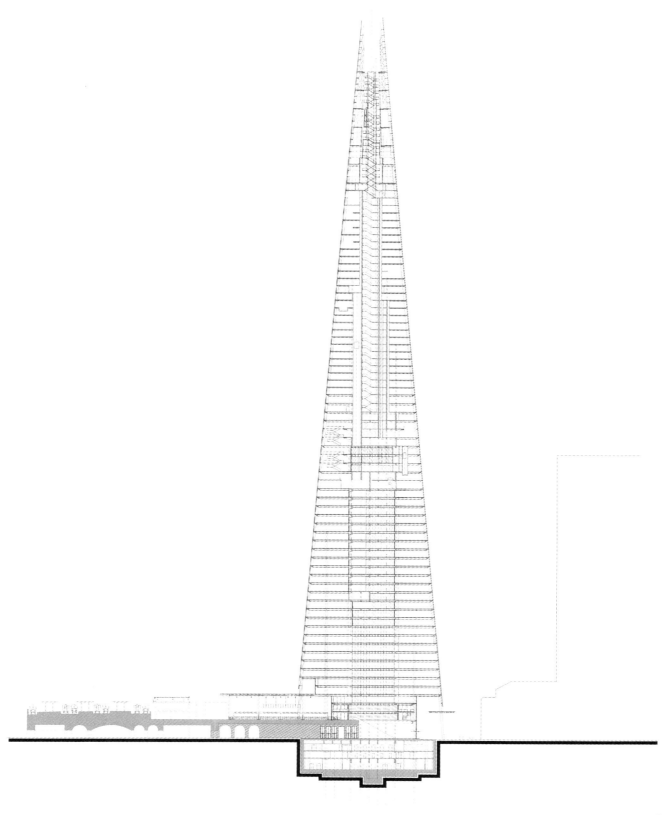

11 剖面

TORRES PORTA FIRA, SPAIN
西班牙费拉 Porta Fira 双子塔

伊东丰雄 | Toyo Ito

项目业主：Hoteles Santos, Realia

建筑设计：Toyo Ito & Associates, Architects+Fermín Vázquez - b720 arquitectos

设计团队：Toyo Ito & Associates, Architects：Toyo Ito, Takeo Higashi, Atsushi Ito, Wataru Fujie, Keisuke Sawamura, Shuichi Kobari, Florian Busch*, Andrew Barrie*, (*ex-staff)

设计团队：Fermín Vázquez（b720 arquitectos）：Fermín Vázquez, Alexa Plasencia*, Cristina Algás, Ana Caffaro*, Laia Isern, Peco Mulet, Pietro Peyron*, Andrea Rodríguez*, Gaëlle Lauxerrois, Magdalena Ostornol*, Mirko Usai*

外部环境合作设计：Kenichi Shinozaki, You Hama

结构体系：IDOM Ingeniería y Sistema S.A.

结构工程设计团队：Joel Montoy, Francis Steiner, Sergio Blanco

机械工程：Grupo JG

机械工程设计团队：Jaume Serra, Germán Romero

建筑顾问：FCC Construcción,S.A./Gerard Alvira, Joaquim Puiggros, Amadeu Portell

设备顾问：Climava/Marc Oliva

  作为巴塞罗那L'Hospitalet市都市再开发项目的一部分，费拉Porta Fira双子塔伫立于欧洲广场内。该基地位于巴塞罗那市和El Prat国际机场之间，距离巴塞罗那El Prat国际机场仅为8 km，并构成了Barcelona Gran Via Fair会场的延伸部分，营造出一个引导人们进入活动区域的门户。

  该工程由3部分组成：高约110 m的宾馆塔楼、办公塔楼以及一个略矮单体。宾馆塔楼外部饰有红色铝管遮光栅格并附着3D式弧线表面。随着建筑高度的升高，建筑体量逐渐延伸并且扭转。办公塔楼采用垂直式透明玻璃箱形结构，内设流线型红色核心筒。两个塔楼在形式上形成鲜明对比，又形成了和谐互补的关系。

  宾馆塔楼共设有345间客房，可为参加交易会及商贸洽谈的访客提供住宿。入口大厅、餐厅以及同时用作会议厅、展厅的宴会厅都位于一层。中层会议厅能够通过滑动墙体分隔成4个空间，这里还没有3个小型会议室。较矮的大楼屋顶上设有一个配有厨房的露台花园，可举行鸡尾酒会。建筑空间相互交融，来宾可以参加鸡尾酒会后，走到一层的宴会厅参加晚宴，并尽情欣赏大楼建筑。

  办公塔楼共设有22个办公层、2个可出租商铺以及1个入口大厅。从入口大厅到塔楼顶层的垂直交通核引导流线。该段垂直交通核象征人体器官。每一层电梯间的开门部分都不相同，在这里我们可以看到紧邻巴塞罗那的Mont Juic山和地中海。在4个租赁部分的办公层，人们可以享受巴塞罗那市的远街和山峦的各种全景。

  近日，城市地铁正式投入使用，使该地区成为连接起都市再开发项目的关键点。"Torres Porta Fira"工程博得了广泛的关注，将成为整个巴塞罗那市和L'Hospitalet市令人瞩目的地标式建筑。

  Within the "European Plaza", a pair of twin towers now stands as part of the urban regeneration project for L'Hospitalet, the neighboring city to Barcelona. The site is approximately 8km away from Barcelona El Prat International Airport towards the Barcelona City direction. With this condition of being in between an airport and the urban area, the project constitutes as part of the extension to the "Barcelona Gran Via Fair" Venue, intending to form a gateway that leads people into the fair. Beneath the energetic red twin towers, the entrance hall we designed spreads out to an area of 240,000 m² and is connected to the continuous circulation corridor, the Central Axis.

  The project consists of three parts; a hotel tower and an office tower, both about 110m tall and a lower compartment which has a roof garden that connects the two towers. The hotel tower has a 3D curved surface clad with red aluminum pipe louvers. The volume gradually expands and twists as it reaches the top where the large hotel suites are located. The office tower on the other hand, has a fluid red core in a transparent rectilinear glass box. As the core is set to one side of the tower, the section of this core can be seen externally. Whilst these two towers have clear contrast in form, they acquire a harmonic and complementary relationship.

  The hotel tower has the capacity of 345 rooms, to provide for expected visitors of trade fairs as well as for business. The entrance hall, restaurant, and the banquet hall which is also used for conferences or exhibitions are located on the ground floor. On the mezzanine level,

there are conference rooms which can be divided into four spaces by sliding partitions and three small meeting rooms. The roof of the lower building volume has a garden terrace that acts as a multifunctional space with a kitchen where cocktail parties can be held. Guests can then migrate to the banquet hall on the ground floor for dinner after the cocktail party, whilst fully appreciating the building as the spaces interlock with each other.

The office tower has 22 office floors, two rentable shops and an entrance hall on the ground floor. The journey from entrance hall to the top level is led by the vertical flows within the red organic form that resembles body organs. The opening sections of elevator halls differ on each floor and from there we can see Mont Juic and the Mediterranean Sea alongside of Barcelona. From an office floor which is able to be divided into four parts for rent, people enjoy the diverse panoramic view of distant streets and mountains of Barcelona.

The city's subway has recently started operating, connecting the major points of the urban regeneration project while the mid-to-high-rise buildings are currently under construction. "Torres Porta Fira" attracts a vast amount of attention as the new landmark is witnessed across Barcelona city and L' Hospitalet.

1 红色的标志性塔楼
2 建筑与环境的融合

3 建筑外观

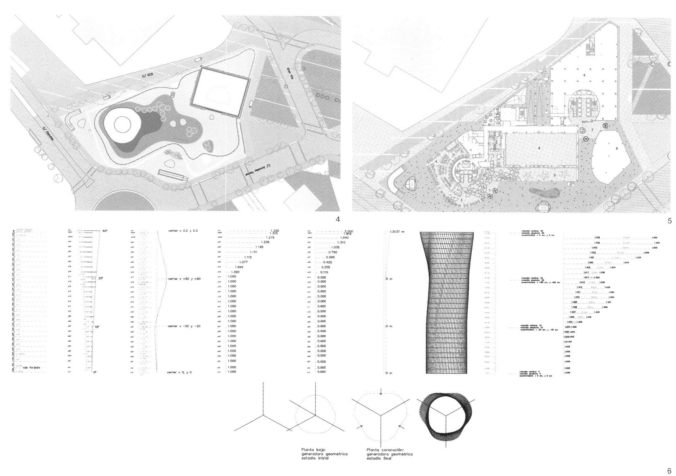

4　总平面
5　一层平面
6　塔楼结构形体生成
7　造型元素的内外统一
8　餐厅实景

HANGZHOU HUALIAN UDG TIME SQUARE
杭州华联 UDG 时代广场

冯·格康，玛格及合伙人建筑师事务所 | gmp

项目名称：华联 UDG 时代广场	合 伙 人：尼古劳斯·格茨
业　　主：华联发展集团	项目负责人：福克玛·西弗斯
建设地点：杭州市江干区海达南路	设计团队：Wiebke Meyenburg, Diana Spanier, Ulrich Rösler, Simone Nentwig, Alexandra Kühne, Barbara Henke
设计单位：gmp	
合作单位：浙江城建设计集团股份有限公司	结构设计：浙江城建设计集团股份有限公司
建筑面积：地上 9.5 万 m²，地下 2.7 万 m²	设备设计：浙江城建设计集团股份有限公司
建筑高度：137 m	照明设计：Schlotfeldt Licht (Berlin)
建筑层数：32 层	建筑设计：2005 年
建筑结构及材料：建筑主体钢筋混凝土，预制单元式幕墙，楼层连接处为金属隔热板	建设周期：2007～2011 年
	图纸版权：gmp
建 筑 师：曼哈德·冯·格康	摄　　影：Hans Georg Esch

华联UDG时代广场建设基地位于杭州市钱塘江边的新中央商务区内，地块直接与区域文化亮点——新建的大剧院相邻。

方案确定为基于相同理念的双塔建筑，因此，设计以四方形作为建筑平面的基本形式，并在此基础上进行多样化造型。

这两座高层建筑都拥有通透和轻盈的外立面。通过精确设计和技术处理的玻璃幕墙赋予建筑现代气质和时代精神，同时，实现了建筑内部面向河岸景观视野的通透。

双塔建筑体块的圆角设计赋予其和谐轻柔的外观形象，之间通过一个3层的裙房相连。双塔偏转对置，以保证各个办公用房拥有最佳的视野并避免相互对视。

外立面的设计赋予双塔建筑个性特征和与众不同的外观。两者相面对一侧的办公楼层设有4层高的阳光大厅，以独一无二的形象成为华联钱江时代广场的"名片"，具备绝佳的广告效应和标志性。

弧线形建筑外墙延伸至室内阳光大厅，形成斜向外壁，并由此生成一个梯形平面。由于这种外向式斜面设计，人们从室内向外观看时拥有一个宽阔的视角。阳光大厅构成室内外之间的"隔热缓冲"空间，有助降低能耗。这一自然通风的气候调节空间在寒冷季节可将热能损失降低至最小，并可利用阳光辐射能为建筑提供保温。大尺寸的进排风口可保证足够的空气流通，防止在夏季建筑内部过热。

就消防设施而言，应将前置的空中花园区作为室外空间对待。通过使用进排风板可保证产生的烟气迅速排出。而中庭的内部立面，即办公用房的外立面，设计为普通玻璃幕墙，玻璃金属结构由金属框材和节能的中空隔热玻璃构成。层间防火封堵使用800 mm高的防火梁加以保证。

外立面采用的呼吸式幕墙是一种混合式解决方案。幕墙结构由封闭式玻璃幕墙和可开启式窗扇两种元素构成，其工作原理和结构并不相同。封闭式玻璃幕墙与空间通高，采用了无色并具有高保温性能的复合安全玻璃，遮阳构件被置于玻璃之后，还可起到防止眩光的作用。可开启式窗扇采用自然通风和遮阳装置相结合的设计，即使在恶劣气候条件下，办公用房内部窗户也可开启进行自然通风，从而大大提高了建筑的舒适性。外层玻璃和内层玻璃之间可进行自然通风，同时装有可升起的遮阳卷帘。整个幕墙结构在预制完成后进行现场组装。

在共用的3层裙房中间处设有大堂、餐厅和若干商店，一个向前伸展的宽大雨篷突出了北侧主入口。

2层的地下停车库设有基地所需的私人停车位以及餐厅和办公区的上货区，并通过一座地下通道实现了与地下停车场的连接。另外，各种设备和防空用房也设置在地下。自行车库可以通过南部的坡道从外面进入。

塔楼的平面布局根据功能要求设计，具有高度的经济实用性。标准层面积约1 494 m²，办公用房的平面设计允许进行多种组合和组织形式，同时可将空间灵活分隔成若干个租售单元。此外，贯通多层的空中花园提高了建筑对使用者的吸引力。

1 建筑实景

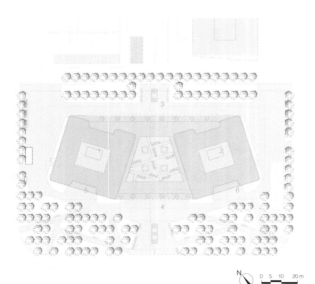

1 A楼
2 B楼
3 入口雨篷
4 主入口
5 办公楼入口
6 次入口

2 双塔建筑体块的圆角设计赋予建筑和谐轻柔的外观形象
3 总平面
4 向前伸展的宽大雨篷突出了北侧主入口
5 一层平面
6 标准层平面

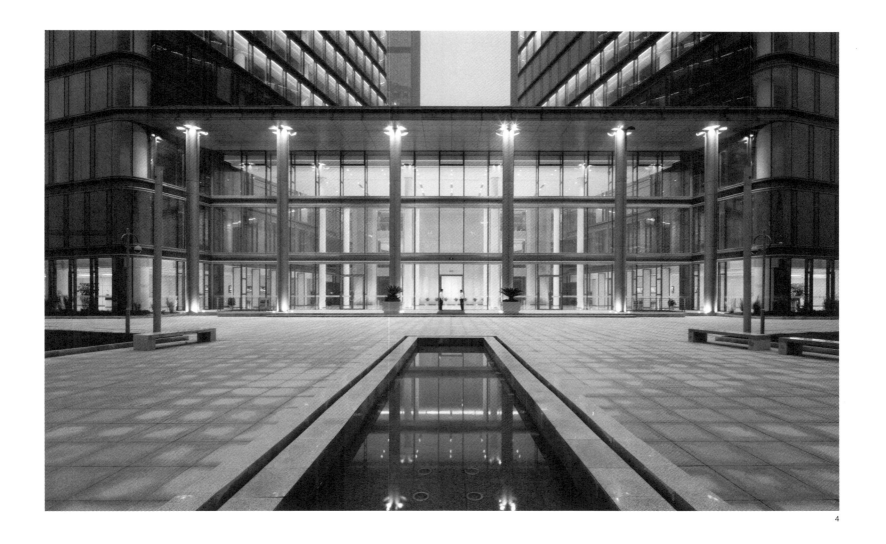

4

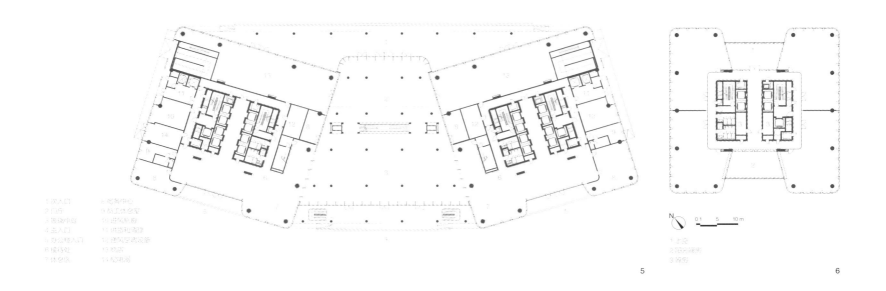

5   6

SÃO PAULO INFINITY TOWER, BRAZIL
巴西圣保罗无限大楼

KPF建筑师事务所 | Kohn Pedersen Fox Associates PC

项目名称：无限大楼
业　　主：YUNY Incorporadora / GTIS Partners
建设地点：巴西圣保罗
设计单位：Kohn Pedersen Fox Associates
设计总负责：William Louie
项目主管：Lloyd Sigal
高级设计师：Jochen Menzer
项目经理：Andrew Cleary，Dominic Dunn
设计团队：Debra Asztalos, Hernaldo Flores，Arthur Tseng，Rachel Villalta

当地建筑师：Aflalo & Gasperini Arquitetos，Vila Olimpia
结构工程师：Aluzio A. M. D´Avila
机械工程师：Teknika Projects and Consulting S/C Ltda.，Sao Paulo
外墙顾问：Israel Berger & Associates, LLC.
景观设计：D/W Santana
LEED 顾问：CTE
图纸版权：Kohn Pedersen Fox Associates
摄　　影：Leonardo Finotti

无限大楼是一座3.9万 m²的办公大厦，坐落于巴西圣保罗市中心，紧邻Juscelino Kubitschek大街与Faria Lima大街的交汇处，处于圣保罗商业区中心的位置确立了其作为城市少有的A级办公建筑之一的战略性地位。

这座被南美洲房地产界誉为"标志性建筑"的大厦优雅地矗立，约120 m的高度超出周围建筑，在圣保罗的天空下展示出独特的魅力。建筑动感的造型、航海的意境，让人不由得联想起航船扬帆前行的种种情景。

"扬帆航行的设计寓意使这座建筑在圣保罗密集的城市环境中彰显出独特并强有力的个性。"项目设计负责人威廉·路易（William Louie）说。

建筑的两个弯曲的玻璃幕墙系统似"帆"雅致地包裹了每一层的内部空间，建筑外部的百叶窗鳍片强调了幕墙系统，并控制着阳光的射入及眩光。外立面的曲率提升了建筑内部的工作环境，并为其中的人们提供了更宽阔的城市全景视野，同时楼层面积也随着高度逐级增大，从而最大化较高楼层的租赁收入。建筑两片幕墙曲线端部设计有挑出的私人露台，提供了北侧Avenida Paulista大街和南侧Faria Lima大街生动的景致。

建筑的底层处设计利用当地温和的气候营造了室内室外化的进入体验，将建筑与城市环境相连接，并提供了从全球第六大人口稠密城市进入建筑内部的安静过渡。倒影池的精致壁毯、带遮阳的步道、园景区域以及带顶篷的车辆门道通向建筑的大堂。大堂以水池环绕，可由两座小桥进入其内。风格雅致的大堂空间由高而弯曲的玻璃界定，墙装饰以丰富的板材，尤以精心挑选的巴西花岗石和二翅豆（Cumaru）木为特色。

建筑的美与其一级的功能相得益彰。与邻近的许多建筑相比，这座18层的大厦实现了当地所能接受的最大建筑面积，同时对各种类型的租户而言更具灵活性与高效性。3 m的楼层净高和高性能玻璃的应用，使建筑获得最大程度的自然采光和视野，除此之外，凭借最先进的技术、地标式的识别性、独特的入口体验以及所有细节的完美品质，无限大楼以同等甚至超越其他国际化城市（如纽约、伦敦和香港）的建筑水准，为巴西确立了新的国际化建筑标准。

KPF的主管Lloyd Sigal补充道："设计一座了不起的建筑是一回事，建设它是另外一回事。我们的客户懂得，对一座建筑的成功以及为巴西市场确立的卓越标准而言，建设质量至关重要。"

1 无限大厦全景

HIGH-RISE BUILDING | DESIGN WORKS

2

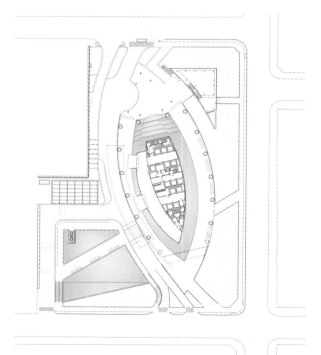

3

Infinity Tower is a 39,000 m² office tower located at the center of São Paulo, Brazil. Its immediate proximity to the intersection of Faria Lima and Juscelino Kubitschek Boulevard, the heart of São Paulo's Financial District, strategically positions the tower as one of the City's only Class A office buildings.

Heralded as an "Architectural Icon" by the South American real estate community, the tower rises elegantly some 120 meters above its surroundings to establish a unique identity against the São Paulo skyline. The dynamic massing of the building, nautical in nature, is reminiscent of a schooner running under full sail.

"The concept of the metaphorical ship tacking to claim its berth on the shores of Faria Lima is a distinct and powerful image within the dense urban context of São Paulo," according William Louie, who led the design of the project.

The building's 'sails' gracefully wrap each level with two sweeping curves of a glass curtainwall system, highlighted by exterior 'brise soleil' fins to control sunlight and glare. The curvature of the facade enhances the work environment and provides broad panoramic views of the city, while each floor plate increases in size as the tower rises to maximize leasing revenue at the upper levels. Private balconies extend

2 入口空间
3 总平面
4 城市与建筑间安静的过渡区

from each end of both curves and afford dramatic views of Avenida Paulista to the north and Faria Lima to the south.

At its base, the building connects to its urban context through an indoor-outdoor entry experience that capitalizes on São Paulo's temperate climate and offers a quiet transition from the world's sixth most populous city. A subtle tapestry of reflecting pools, shaded walkways, landscaped areas and a canopy covered porte-cochere lead to the building's main lobby, which is surrounded by water and accessed across two bridges. The lobby itself is gracefully defined by a high, curving glass wall and rich material palette featuring carefully selected Brazilian granites and Cumaru wood.

The building's beauty is matched by its world-class functionality. In contrast to many of its neighbors, the 18-story tower achieves the maximum floor area that São Paulo allows, while also providing flexibility and efficiency for a variety of tenant types. With three-meter floor-to-ceiling heights and high-performance glass that maximize access to natural light and views, state-of-the-art technology, a landmark identity, a distinct entry experience, and quality of detail and finish at every scale, Infinity Tower sets a new benchmark for the Brazilian market with international building standards matching or exceeding those of other global cities such as New York, London, and Hong Kong.

KPF Managing Principal, Lloyd Sigal, adds, "It is one thing to design a great building. It is another to build it. Our client understood that quality of execution was vital to the building's success and defining a new standard for excellence in the Brazilian market."

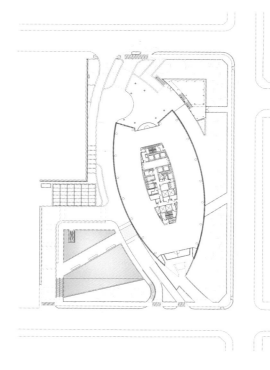

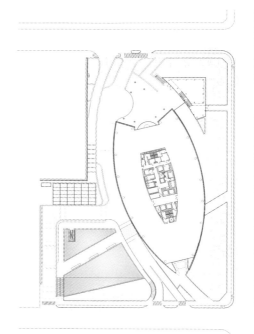

5 办公大厦入口区
6 低层区标准层
7 高层区标准层

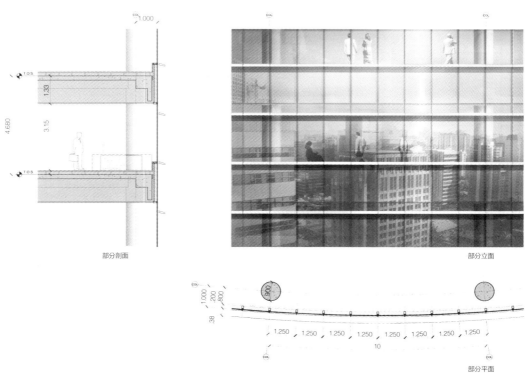

部分剖面　　部分立面

部分平面

8　弯曲的玻璃幕墙
9　幕墙细部构造

ABU DHABI INVESTMENT AUTHORITY HEADQUARTERS, UAE
阿拉伯联合酋长国阿布扎比投资管理局总部大厦

KPF建筑师事务所 | Kohn Pedersen Fox Associates PC

项目业主：阿布扎比投资管理局
设备设施：Corporate Headquarters
项目规模：87 300 m²
项目状态：2007年竣工
所获奖项：2008年中东房地产最佳商业/办公建筑楼盘奖
2008年ULI欧洲区杰出建筑奖项入围奖
2008年美国芝加哥Anthenaeum国际建筑奖
2008年海湾国家竣工年度办公/商业建筑奖

一座气派非凡的总部大厦乃是一家企业甚至整座城市的象征。它应汇聚众人，同时营造出令人舒适、愉悦而又催人奋进的工作环境。一座全新的总部大楼的设计亦是塑造办公空间，以反映出企业文化与愿景的契机。

我们计划根据伊斯兰文化传统，设计出庭院式的空间造型。大厦由分为两翼的条状体量构成，中间由垂直天井连接。经过仔细研究，我们发现阿布扎比这座城市具有突出而典型的棋格式街道布局，规划方案恰恰反映了这一特点：楼体北翼延续了棋格样式，南翼则形似一本书，向着海洋的美好风光以及圣城麦加敞开。开放的空间将海景以及滨海大街的碧绿景致收进楼内；而一系列空中花园则成为了绿荫掩映的公园街道的延伸。

建筑设计的关键在于表现海洋以及城市规划方案在整块建筑基址开发过程中的深远影响。该设计方案风格简约、明快。阿布扎比投资管理局（ADIA）要求设计大型的完整空间，根据当地议会传统，每层应设有宽敞的中央公共非正式集会和社交场地。中央天井花园正好就是这样一个区域，它成为了整个设计方案的核心所在，将整个建筑联系、统一起来，体现了阿布扎比投资管理局的组织结构以及透明度。这一设计理念更深层次体现了本地化造型设计的创意思想，这一创意的灵感来源正是这一滨水地带所具有的特色以及建筑本身作为国际化公司总部大厦的非凡意义。38层的大厦提供了总计87 300 m²的空间面积，其中包括62 500 m²的办公空间。ADIA强烈要求其全新的总部大楼能够体现透明、合作的工作精神。天井内的中央会议区毗邻每层的电梯间，此布局紧扣设计主题，同时成为满足甲方要求的解决方案。写字间分列天井两侧，从而建立起办公室同中央会议区的视觉联系。

服务区位于楼内东部边缘，两翼写字间空间可以灵活调整。各部门执行总监办公室设于各翼办公区的首要位置，其对面是部门会议套间。会议套间的设置促进了企业员工的互动交流，体现了合作、交流以及透明办公的企业文化理念。

灵活的空间配置在办公空间楼层规划当中是必不可少的。专为隔断、顶棚、服务设施以及办公设备开发设计的组合系统能够适应高度多变的办公环境要求，同时满足各部门的不同需要。互动会议区同开放的组织规划以及蜂窝式的写字间一样至关重要。会议区设置于中央天井景观带及一系列空中花园之中。天井覆面系统融合运用了活动式百叶窗结构，在保证公共区域免受西面日晒之苦的同时保证了外界的景观视野。建筑使用者可以通过手持控制器独立控制办公环境（包括照明、空调以及外部百叶窗等系统），并且各系统位置方便可寻，随时可根据办公室布局变化做出调整。

此外，写字间还应满足该跨国公司区域工作的不同工作时间的要求。设计方案为满足24小时生活及工作所需，专门设置、整合了咖啡厅、餐厅、祈祷室、健身俱乐部及游泳池等辅助设施。

建筑设计做到了有效利用各类自然资源。自然光照明通过全玻璃立面以及磨砂玻璃隔断引入各办公室；伴随全玻璃立面产生的诸如阳光调控及热量舒适度调节等问题最终通过"活动"立面得以解决；"活动"立面包括3层：外部为低辐射（率）双层镀膜玻璃单元，内部为200 mm中空多孔百叶窗结构，以及独立的内部可操控玻璃单元。多孔结构由楼内管理系统控制，日光直射时窗体完全关闭，避免了过度日照。同时，办公室的空气通过孔结构导出，排除余热，阻止了玻璃墙体由于日照而在办公室内汇聚过多的热量。这一整套体系为建筑的使用者提供了高度舒适的温控系统。

The creation of a corporate headquarters is an important moment in the life of a company. A great headquarters becomes both a symbol of the company and the symbol of its city. A headquarters building should bring people together and create a work environment that is comfortable, pleasant and stimulating. In the design of a new headquarters there is the opportunity to shape the workplace in a way that reflects the aspirations and culture of the company.

To begin, we thought of a space that would be like a courtyard, in the tradition of Islamic architecture. Here the courtyard is formed by two bars set apart as wings connected by a central vertical courtyard atrium. We examined the city of Abu Dhabi: its strong urban grid. The plan reflects this. The wing to the north follows the city grid while the wing to the south appears to open like a book, opening to the sea, the vista and towards Mecca. The opening draws the sea and green of the Corniche into the building. A series of gardens in the sky become an extension of the green parkway.

The key to the design is the acknowledgement of the profound importance of the sea in the development of the site and of the urban plan as a garden city. The scheme is simple. ADIA's future requirements demand a large single floor plate with a large central common zone on each floor for informal meeting and social interaction in the tradition of the local majlis. This central zone has become the vertical atrium garden, the heart of the scheme that unifies the whole building and represents the Abu Dhabi Investment Authority organization and its openness. The design concept further refines the idea of creating an indigenous form, a form inspired by the special character of this waterfront site and the buildings importance as an international headquarters. The 38 storey tower provides a total of 87,300 square metres of accommodation and support space, inclusive of 62,500 square metres of office space. ADIA was keen that their new headquarters promoted a new spirit of transparency and cooperative working. The central meeting area in the atrium, adjacent to the lift lobby on each floor, made an ideal focus for the scheme and also gave an architectural solution that matched the organization's aspirations – the office wings each side of the atrium have been designed so that there is a visual connection with the central meeting area, to give the sense of a common space dedicated to a common purpose.

The service core anchors the eastern edge of the building al-

1 滨水夜景 (Image Credit H. G. Esch)

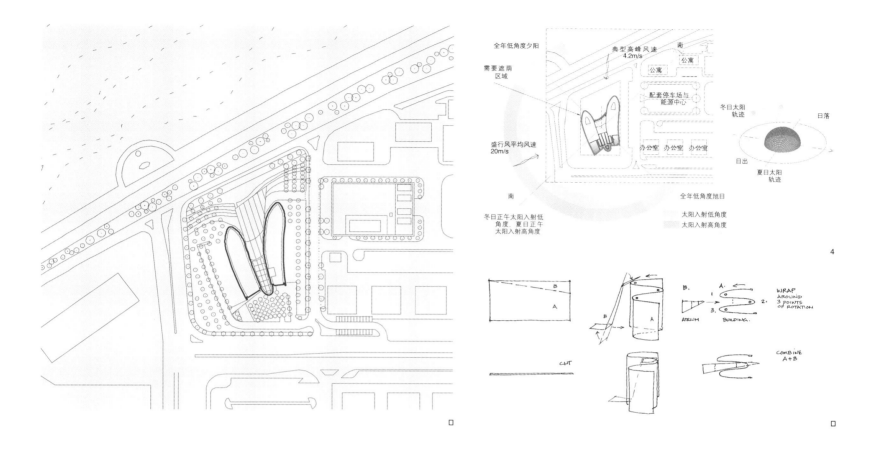

2　建筑实景 (Zmage Credit H. G. Esch)
3　总平面
4　日照采光分析
5　形态构思

lowing flexible workspace in the two office wings. At the head of one office wing is the office of the Executive Director of each department; at the head of the opposite wing is the large departmental conference suite. Their placement encourages interaction between staff members during the day, reinforcing the ethos of cooperation, communication and transparency.

Flexibility was a necessity in the planning of the office floors; a high churn rate was anticipated and designed for. Modular systems for partitions, ceilings, services and furniture systems were developed to accommodate the high rate of change, while also providing for the various requirements of each specific department. Open-plan and cellular office spaces were required, as were zones for interaction and meetings. The latter are located both within a central landscaped atrium and the series of sky courtyard gardens. The atria cladding system incorporates activated fabric blinds, sheltering public spaces from the western sun while providing views out. The building user has control of his individual office environment (lighting, comfort cooling and external blinds) through a hand-held control unit, and each of the building systems are addressable, and can be reconfigured to suit changes to the office layout.

The office space also needed to accommodate different working hours associated with this global company's regional working needs. The design caters to this notion of the 24/7 live and work concept with café/restaurant, prayer rooms, health club and swimming pool amenities incorporated into the design.

The building is alive and responsive to natural forces. Natural daylight is provided to the office interiors through a fully glazed facade, and by frosted-glazed office partitions. With a fully glazed facade issues of solar control and thermal comfort are handled using an 'active' facade. The facade comprises of three layers: a low-e-coated outer double glazed unit, a cavity of 200mm inside which is a perforated solar control blind, and a single inner glazed operable unit. The blinds are controlled by the building management system, so the blind is closed when the facade is in direct sun, preventing solar gain; at the same time air from the office is drawn through the cavity and extracted, removing any excess heat, and preventing heat from radiating through the facade into the office area. These systems provide a high level of thermal comfort for the building user.

HIGH-RISE BUILDING | DESIGN WORKS

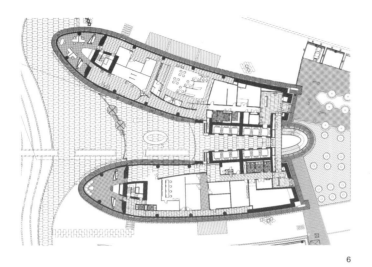

6

7

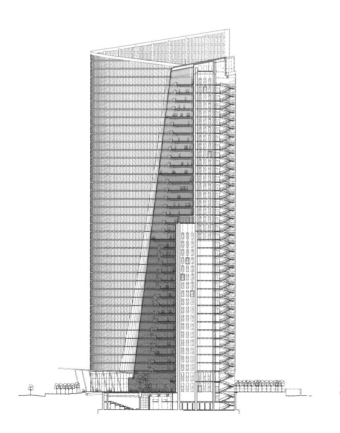

8

9

10

11

12

6　首层平面
7　三层平面
8　剖面图
9　立面图
10　景观朝向俱佳的办公空间
　　(Image Credit H. G. Esch)
11　流畅连续的休息空间
　　(Image Credit H. G. Esch)
12　层次丰富的交通空间
　　(Image Credit H. G. Esch)

HIGH-RISE BUILDING | DESIGN WORKS

SAMSUNG SEOCHO, KOREA
韩国三星瑞草大厦

KPF建筑师事务所 | Kohn Pedersen Fox Associates PC

项目业主：三星电子，三星C&T，三星人寿保险
项目功能：企业办公
机电设备：合作设计
建筑面积：23.9万 m^2
项目状态：竣工

三星瑞草大厦的设计方案打造了一座高中低层紧密联系、同首尔市中心周边城市环境相整合的办公中心，这种整合方式同久负盛名的洛克菲勒中心与纽约的整合颇为相似。

受代表韩国传统木艺水平的细木工手艺启发，楼体采用了互相接合的块状造型，意在将建筑方案独特的功能模块互相连接整合为一体。将一座典型韩国"超级大厦"分解成互相联系的元素，各部分彼此呼应，促进基址内的人行交通，同时与周边城市环境紧密衔接，加强了同既有建筑的联系。

为进一步明确各建筑体块的功能划分，垂直和水平方向的外覆材质采用两种不同直棂方案，从而凸显各楼体间的相互关系。大楼紧密相连的体块间所产生的反射由装饰栅隔挡，满足了大厦内各层小型器械室允许使用高效的补充边缘空调系统的特殊要求。水平与垂直幕墙的对比以及反射玻璃的使用也强化了悬臂式互锁塔状造型的戏剧性艺术效果。而给人以自由浮动之感的体块设计也得益于装饰栅隔挡了相互反射。

两类幕墙间的狭窄凹形空间采用半透明玻璃覆面，而面积较大的立面区域仍由传统中空玻璃单元（IGUs）覆盖。透过观景玻璃的光通量由遮阳系统自动控制，而各层凹形空间均配有可控窗体。这套综合幕墙系统为使用者提供了最自由的照明及自然通风控制，将建筑对人造光源、供暖以及制冷系统的依赖性降至最低。

三星瑞草大厦总建筑面积为238 560 m^2，塔楼A座和B座为三星集团以及其他供应商、制造商提供了办公空间，而C座则成为三星电子公司的全新总部大楼。大厦C座根据功能规划分为多个体块，餐饮、会议以及演播功能融入其中，同时作为三星展览中心承载着展示三星电子最新科研成果的任务。

在大厦的景观设计中，方案规划出串连综合公共区域的人行道路，令基址内步行活动更为便利。塔楼A座与B座间设有公司公园，并且设置横穿整个场地的带有顶棚的廊道直通C座。另外，大型公共中心可放映电影，举行音乐会及开展其他各类活动，从而进一步促进楼内各公司社交活动的开展。

The design for the Samsung Seocho project creates an office complex where linkages are created at the low- mid and high-rise levels to fully integrate the project within the surrounding urban context of downtown Seoul in the same manner that the famed Rockefeller Center accomplishes this in New York City.

Inspired by the symbolic joinery represented in traditional Korean woodworking, the massing of the buildings uses interlocking forms that aim to inter-weave the building's distinct program elements. Taking a typical Korean "superblock" and breaking it down into inter-related elements, the buildings gesture in towards each other to encourage pedestrian movement and interaction throughout the site, and out to the surrounding city to facilitate linkages with existing buildings.

To enhance the clarity of individual building volumes, vertical and horizontal textures were introduced to each of the surfaces by using two distinct mullion systems, thereby heightening the inter-relationship between the buildings. The reveals between the intertwined volumes of the towers are louvered, accommodating an unusual project

1 建筑实景 (Image Credit Jasung E)

HIGH-RISE BUILDING | DESIGN WORKS

2

2 临街入口
(Image Credit H. G. Esch)
3 塔楼间的绿化空间
(Image Credit H. G. Esch)
4 屋顶景观空间
(Image Credit JaesungE)
5 底层入口空间
(Image Credit JaesungE)

requirement for small mechanical rooms on every floor that allow for an efficient, supplemental perimeter cooling system. The contrast of horizontal and vertical curtain wall types and the use of reflective glass reinforce the drama of the interlocking, cantilevered tower masses, which seem to float free of each other, thanks to the louvered reveals.

For both curtain wall types, the narrow recess has translucent glazing, while the broader areas are glazed with traditional vision insulated glazing units (IGUs). Automated shades control the amount of light allowed through the vision glazing, while the recesses feature operable windows on all floors. These complex curtain wall systems allow for maximal control by the occupants over daylighting and natural ventilation, minimizing the amount of artificial lighting, heating and cooling required by the buildings.

Encompassing a total built-up area of 238,560 square meters, the Samsung Seocho project provides office space for the Samsung Group and various suppliers/manufacturers in Towers A and B, while Tower C serves as the new corporate headquarters for Samsung Electronics. The Tower C podium is broken into multiple program-specific volumes to establish a human scale for the project. Dining, meeting and exhibition functions are accommodated in several masses, with the most important being the Samsung Showroom showcasing the latest Samsung technologies.

The landscape plan creates paved walkways that unify and traverse the complexities of the public areas, facilitating pedestrian movement throughout the site: a corporate park between Towers A and B, through-block pedestrian breezeways in the podium of Tower C, and a large public plaza, which allow for public movie screenings, concerts and other events to further encourage social interaction on-site.

HIGH-RISE BUILDING | DESIGN WORKS

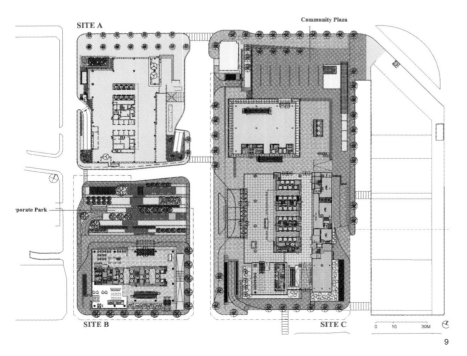

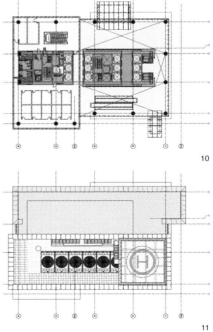

6 剖面
7 结构受力分析
8 机械通风分析
9 首层平面
10 B栋二层平面
11 B栋顶层平面

12 玻璃幕墙界面 (Image Credit H. G. Esch)
13 休闲空间一隅 (Image Credit H. G. Esch)
14 开敞的交流空间 (Image Credit H. G. Esch)
15 交通空间 (Image Credit H. G. Esch)
16 私密的会谈空间 (Image Credit H. G. Esch)

THE PINNACLE, ENGLAND
英国尖塔

KPF建筑师事务所 | Kohn Pedersen Fox Architects PC

项目委托：尖塔第一有限公司（The Pinnacle No.1 Ltd）
项目设施：办公建筑
项目规模：8.8万 m²

坐落于主教门和克洛斯比广场的尖塔又名主教门大厦（The Bishopsgate Tower），它即将成为城市中最重要的新建筑之一，增强此区域内高层建筑群的整体特征和可识别性。该项目同时也将为公共空间做出实质性的贡献——为行人打通地面区域，并联结若干沿主教门和圣玛丽斧街的重要城市空间。

建筑物的几何形状由简单剪切的锥体嵌入镶边的逐渐缩减的平面构成。遮篷则由一系列沿立面和剪切锥体边缘展开的相切弧形构成，并由边缘的一条法曲线控制曲率。

可持续系统是建筑设计的重要组成部分。符合空气动力学的流线外形改善了建筑立面自然通风的性能。玻璃立面保证了充足的自然采光，减少了人工照明的需求量，同时保护了遮阳设备。光电电池进一步减少了能源消耗。

The Pinnacle (formerly known as The Bishopsgate Tower), located at Bishopsgate and Crosby Square will become one of the most significant new buildings in the City with a design that will strengthen the overall character and identity of the emerging cluster of tall buildings in this location. The proposal would also make a substantial contribution to the publicrealm, opening up the ground level area to pedestrians and linking a number of important urban spaces along Bishopsgate and St. Mary Axe.

The building's geometry is composed of simple sheared cones filleting to tapered planes, creating the design surface. The canopy is built from a set of tangential arcs unwrapped parametrically along the facade planes and sheared cones at its edge controlled by a law curve defining the curvature in space of the canopy edge.

Sustainable systems are integral to the architectural design. The aerodynamic shape improves the performance of the naturally ventilating facade with its snakeskin design. The outer layer of glass protects the sun-shading which reduces heat gain. The glass facade allows ample natural light, reducing the amount of artificial light required. Photovoltaic cells further reduce energy consumption.

1 伦敦尖塔实景（Credit KPF and Cityscape）
2 办公标准层平面

1

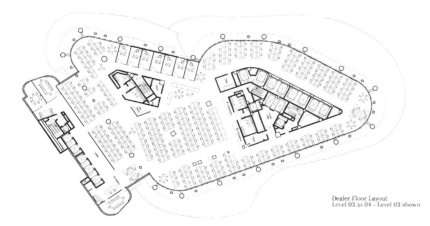

Dealer Floor Layout
Level 03 to 04 - Level 03 shown

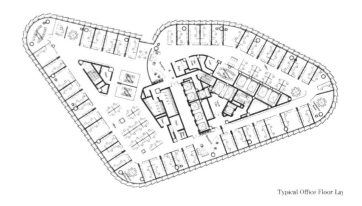

Typical Office Floor Lay

2

HIGH-RISE BUILDING | DESIGN WORKS

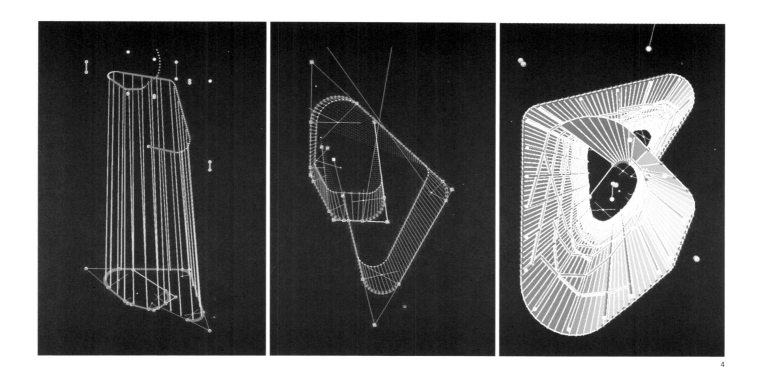

4

5

3 从圣保罗教堂看伦敦尖塔（Credit Rendering: Cityscape）
4 建筑形体生成分析
5 建筑形体生成与街区环境肌理

NORTHEAST ASIA TRADE TOWER, KOREA
韩国东北亚贸易大厦

KPF建筑师事务所 | Kohn Pedersen Fox Architects PC

项目业主：NSC（Gale International and Posco E&C）
项目类型：办公、居住、酒店、零售和娱乐综合体
项目规模：1 500 000 gsf
联合设计：Heerim Architects & Engineers

东北亚贸易大厦坐落在南国西海岸，位于黄海水平面306 m 之上，可俯瞰整座松岛国际城。项目基地位于公园林荫道与会议中心大道交汇处，交通便利。由于新建了立交桥，从这里开车到机场只需15分钟。大厦的落成不仅是这座新兴国际自由贸易区的第一栋商业大楼，还会成为吸引世界和韩国顶级公司进驻的地标，从而提升人们的生活质量，塑造出松岛国际城的中心性建筑。

大厦建筑面积约为16.7万 m²，共70层，包括商用办公室、服务型公寓、酒店、零售空间和地下停车场等部分，同时还设有五星级住宿空间以及可眺望黄海的观景台，从顶级餐厅和酒吧均可俯瞰地面主广场。

大厦办公区域和服务型公寓的入口分别设在平面的不同方向，1～27层为商用办公区域，面积为6.67万 m²；34层是大型的空中休闲俱乐部，包含餐馆、会议室、健身俱乐部、SPA 疗养室和游泳池；35～63层是为高级管理人员提供的165个高端服务型公寓，从塔楼顶层的观光台远眺，人们可一览新城广阔的景致。

大楼外部通向一个零售和购物中心，这里有很多著名的国际品牌旗舰店。服务型公寓布置在这里，不仅可以饱览中央公园、黄海和仁川城的全景，还会吸引有发展远景的住户。

Rising 1,005 feet above the yellow sea and overlooking Songdo International City on the west coast of South Korea. Located on the corner of Park Avenue and the Convention Center Way the Asia Trade Tower is easily accessible. The new bridge will reduce the driving time to the airport to under 15 minutes. The Northeast Asia Trade Tower will be the first of many commercial buildings within a newly created international free trade zone. The tower complex will be the center piece for Songdo International City and will provide easy access to all associated "quality of life" components.

The new 1.5 million SF tower will provide a landmark location for the very top tier Global and Korean companies as well as providing 5 star living and an observations deck with a view over the Yellow Sea. It will include commercial office, service apartment, hotel, retail space and underground parking and will stand 70 stories tall.

The office entry and the service apartment entrances are separated on different facades, and an exclusive restaurant and bar will overlook the grand plaza at ground level. Approximately 600,000 GSF of commercial office will be located in floors 1~27 and about 165 high-end serviced apartments aimed towards high-end executives will occupy floors 35~63. A sky lounge club on the 34th floor will include a restaurant, meeting space, a health club and spa, and a pool, and an observation deck on the tower's top floor will offer sweeping vistas.

The tower is connected to a retail and shopping destination which will be home to the flagship boutiques of many famous international fashion names. With panoramic views of Central Park, the Yellow Sea and the city of Inch eon, the service apartments will be fully equipped to appeal to discriminating tastes.

1 整体景观

HIGH-RISE BUILDING | DESIGN WORKS

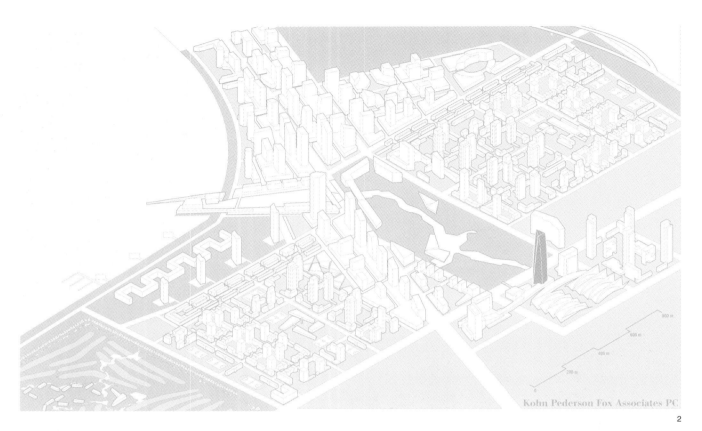

2 区位图
3 入口景观

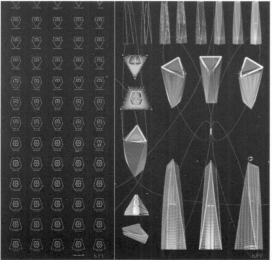

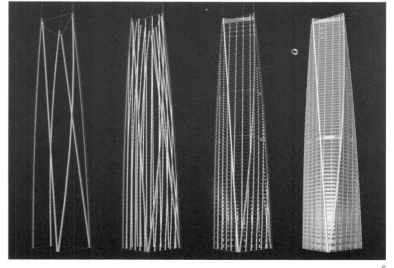

4 夜景鸟瞰
5 平面分析及形体生成
6 建筑形体分析

SONGDO INTERNATIONAL CITY, SOUTH KOREA
韩国松岛国际城住宅

KPF建筑师事务所 | Kohn Pedersen Fox Architects PC

项目业主：NSC (gale international and Posco e&c)
项目功能：居住、办公、零售、娱乐
项目规模：3.7 million gsf, 2 545套居住单元
项目状态：委托设计建设中
项目获奖：2005, AIA New York awards chapter

松岛国际城住宅区位于松岛国际城东部125地块，其严谨的布局形式提升了整体规划的私密性。在规划中，住区形成了一个自组织、自循环的空间结构框架，极富特色。住区中心有一大片湖面，优美、宜人的滨水空间营造，使人如同置身于公园，而位于水面中心的木质茶室面积虽小，却含意丰富，它的存在使整个设计变得厚重起来。

这个工程被划分为4个小区域，每个区域都拥有复合性功能特征，如在住宅楼下都配有零售、住宿和工作空间。在其中的两个区域内，零售空间又被分配给家具设计店和大型商店，剩下的空间则是小规模的精品店、咖啡店和生活馆。

在住区中，每栋住宅楼似乎都力争滨水而建，它们之间仅靠一块伸向水面的木平台连接，而零售和办公住宅空间则围合出街道的边缘形态。这就使住宅单元能在两个方向上获得大量的光照，其景观视野也如松岛城一样多姿多彩。其中，办公住宅单元被设计成单层或双层的阁楼组合形式，通过内部中厅和下面的健身房与零售店相连。建筑立面设计如实地表达出建筑内部的空间属性。大量的玻璃幕墙和金属材质的使用，实现了建筑对周围环境的开放，而木质材料的应用则营造出一种温馨的氛围。

Set at the east edge of Songdo International City, block 125 has been carefully arranged to enhance the urban rigor of the established master plan. This allows the block to maintain a protective frame around its own, more peaceful internal identity. With a large reflecting pool at its heart, surrounded by harmonious, park-like spaces, this development is brought more solemnity with the small but significant placement of a wooden teahouse at the center.

The project has been broken down into four smaller blocks, or quadrants, each block will accommodate a mix of uses, with retail and officetel live/work spaces set below larger residential towers. On two of the blocks, the retail space is allocated for a large home design store and another mega-store. The remaining retail spaces are likely smaller boutiques, coffee shops, or lifestyle stores.

Each of the residential towers appears to float tangentially along the water's edge, connected simply by a wooden deck that juts out into the water, while the retail and officetel form the street edges. This allows the residential units generous daylight, often from two directions, and views as varied as the city of Songdo itself. The officetel units are conceived as a combination of single and double height loftlike spaces, connected by internal atriums to health club facilities and retail shopping below. Architecturally, each of these uses finds expression on the elevations of the building volumes. Lots of glass and metal allows for a more open experience of the surroundings, while wood provides a warm accent.

1 住宅内部景观效果

HIGH-RISE BUILDING | DESIGN WORKS

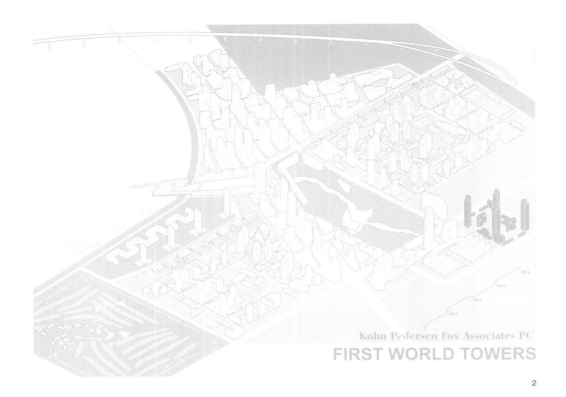

2

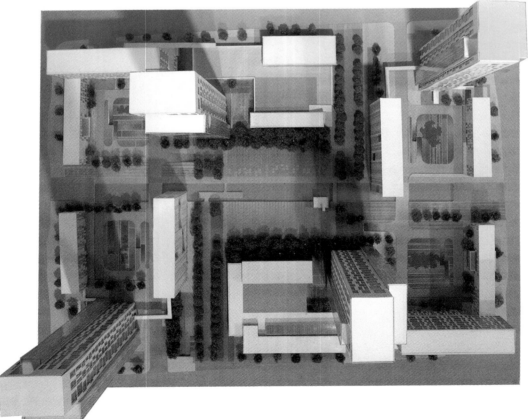

3

2　区位图
3　总平面
4　临街立面
5　整体夜景效果
6　住宅内部建筑景观
7　住宅内部建筑景观

4

5

6

7

JOCKEY CLUB INNOVATION TOWER, HONG KONG POLYTECHNIC UNIVERSITY
香港理工大学赛马会创新楼

扎哈·哈迪德建筑师事务所 | Zaha Hadid Architects

项目名称：香港理工大学赛马会创新楼
业　　主：香港理工大学（校园发展处）
建设地点：11 Yuk Choi Road, Block V, Hong Kong Polytechnic University, Hung Hom, Kowloon, Hong Kong
用地面积：0.66 hm²
建筑面积：2.8万 m²
建筑层数：地上15层，地下1层
建筑高度：78 m
设计单位：Zaha Hadid Architects
当地设计单位：AGC Design Ltd. (Hong Kong), AD+RG（竞赛阶段）
设计总负责：Zaha Hadid, Patrik Schumacher
项目主管（ZHA Project Director）：Woody K.T. Yao
项目负责人（ZHA Project Leader）：Simon K.M. Yu
建筑专业：Hinki Kwong, Melodie Leung, Long Jiang, Zhenjiang Guo, Yang Jingwen, Miron Mutyaba, Pavlos Xanthopoulus, Margarita Yordanova Valova（竞赛阶段）
　　　　　Hinki Kwong, Jinqi Huang, Bianca Cheung, Charles Kwan, Juan Liu, Junkai Jian, Zhenjiang Guo, Uli Blum, Long Jiang, Yang Jingwen, Bessie Tam, Koren Sin, Xu Hui, Tian Zhong（项目阶段）
顾问团队：Ove Arup & Partners, Hong Kong Ltd.（结构和岩土工程，建筑设备工程），Ove Arup & Partners, Hong Kong Ltd.（外墙），Team 73 Hong Kong Ltd.（景观），Westwood Hong & Associates Ltd.（声学），Rider Levett Bucknall Ltd.（工料测量）
施工单位：Shui On Construction Company Ltd., Hong Kong（总承包商），YKK AP Facade Hong Kong LTD / Beijing Jangho Curtain Wall Co., Ltd.（外墙承包商）
设计时间：2007年
建成时间：2013年
图纸版权：Zaha Hadid Architects
摄　　影：Iwan Baan, Doublespace, Virgile Simon Bertrand

　　赛马会创新楼是香港理工大学设计学院和赛马会社会创新设计院的总部。建筑地上15层，使用面积1.5万 m²，可容纳1 800多名师生，楼内设有用于设计教学及创新的各种空间，包括设计工作室、实验室和创作室、展示区、多功能教室、报告厅和公共空间。

　　香港理工大学校园的规划建设持续了半个多世纪，不同学院的教学科研楼风格迥异，但在视觉上又相互协调。赛马会创新楼创造了一种全新的城市空间，丰富了多样的大学生活，同时也展现出设计学院面向未来的发展活力。

　　建筑坐落在校园东北角的一块狭窄且不规则的场地内（南临大学足球场，北临Chatham路和九龙高架高速路的交叉口），与校园中心紧密相连，促进了香港理工大学各个学院与社会团体、政府、企业、NGO（非政府组织）和研究所共同开展多学科领域的合作。

　　赛马会创新楼以更具流动性的组合解构了典型的塔楼/裙房造型。室内外庭院营造了用于会面和交流的非正式空间，与大型展场、工作室、剧场和娱乐设施相映成趣。通过将设计学院的各种课程相关联，创新楼的设计创造了一个多学科共融的环境，同时也建立了一种创作和创新相互启迪的集体研究文化。

　　学生、教师和来访者可以在全部楼层的学院工作室、创作室、实验室、展示区和活动区自由出入。室内的玻璃隔断和中庭空间使各部分空间彼此通透，精心设计的环形交通和公共空间激发了众多学习集群和设计学科间的交流。

　　香港赛马会慈善信托基金为赛马会创新楼投资了2.49亿元港币，同时也为赛马会社会创新设计院的发展提供基金。

　　工程师们应用三维模型和BIM解决赛马会创新楼层层渐变的几何形式。奥雅纳工程顾问香港有限公司（Ove Arup & Partners, Hong Kong Ltd.）负责结构和设备设计，使扎哈·哈迪德建筑师事务所的设计方案变成现实。BIM对这个三维不规则造型建筑的设计和建造而言，是不可或缺的，可以进行建筑构件和设备管道的碰撞分析，同时也可为建筑、结构、设备和外墙设计工程师提供一种有效的沟通模型。奥雅纳公司应用BIM将设备设计与该建筑独特的内部几何形态结合在一起，这在传统的二维图纸中是不可能被完全表达出来的。

　　"外墙是这座建筑的关键要素之一。每周设计讨论会最受关注的重点通常是核对通过现场测量结构构件、建筑设备和外墙结构建立的BIM信息与建筑师提供的设计信息是否一致，"奥雅纳公司的发言人谈道，"BIM经常能够凸显一些顾问和承包商用传统的投影作图法无法发现的设计问题，在BIM的帮助下，各专业间的协调更加有效，也避免了许多现场的错误施工。"

　　同时，奥雅纳公司还应用Etabs软件进行结构分析。在Etabs模型的协助下,工程师们生成了与扎哈·哈迪德建筑师事务所三维计算机模型相匹配的结构轮廓。

创新楼的结构设计难点包括悬挑于人行道之上的塔楼和场地北侧倾斜的立面,因为人行道上不允许埋设建筑基础。为了解决这些难题,奥雅纳公司设计的上部结构采用了3个核心筒和梁柱框架以支撑侧向荷载和建筑的偏心荷载。有些区域采用了斜柱以承受倾斜荷载。为了尽可能地解放建筑一层和二层的空间,三层采用了不连续的转换梁。

塔楼的基础采用了54个大口径钻孔灌注桩可以使结构振动和地面沉降对周围建筑的影响最小化,并且增强了抵抗塔楼倾斜所带来巨大倾斜荷载的承载力。在施工过程中,为了防止周边地面下陷,奥雅纳公司在开挖地下室时设置了报警器监测水平面。

建筑拥有独特的几何形态,为了通过弯曲的三维外墙实现扎哈·哈迪德建筑师事务所的设计意图,奥雅纳公司测试了多种表皮材料,包括纤维增强塑料、张拉材料、铝,最终选择了一种独特的三维金属面板。多层面板的最外层的连接处是开放的,而其下隐藏的一层面板具有遮挡风雨及排水功能。

该项目外墙设计最大的挑战在于解决弯曲三维外墙面板细部和交接处的构造。表皮后面设有若干条狭窄的过道,可以通过可移动的面板进入建筑维修单位在正常维护时无法触及的室内和玻璃幕墙。设置面板的空间可以在建筑开放时由设备管理人员安全关闭。

为了获得稳定的水压和便于将来的改造,奥雅纳公司在冷水系统中应用了不同的压力调节器。冷水系统的环路设计可以根据香港理工大学屋宇设备工程学系Francis Yik教授在一篇科技论文中研究的方法测量冷水机组的满负荷或部分负荷情况。

创新楼的建筑设备采取了多种节能的做法,包括使用水冷式冷水机组的蒸发式冷却塔。新风机组的能量交换器可在废气与新风交换时回收能量,综合教学用房和办公室可以根据CO_2探测器(反馈的数值)调整新鲜空气的供应量。

嵌入式恒温器通过探测其设备内空气样本的温度来获取房间内的平均温度值。室内还安装了可代替回转式热交换器的能量湿度回收通风机以回收实际的和潜在的废热。中央控制和监测系统管理着所有的照明和机械设备以便最大化地节约能源。

1 赛马会创新楼创造了一种全新的城市空间

2 建筑南临大学足球场，坐落在校园内的一块狭窄且不规则的场地中，与校园中心紧密相连

The Jockey Club Innovation Tower (JCIT) is home to the Hong Kong Polytechnic University (PolyU) School of Design, and the Jockey Club Design Institute for Social Innovation. The 15-storey, 15,000 sq. m. tower accommodates more than 1,800 students and staff, with facilities for design education and innovation that include: design studios, labs and workshops, exhibition areas, multi-functional classrooms, lecture theatre and communal lounge.

The Hong Kong Polytechnic University campus has developed its urban fabric over the last 50 years with the university's many faculties housed in visually coherent, yet very different buildings. The JCIT creates a new urban space that enriches the diversity of university life and expresses the dynamism of an institution looking to the future.

Located on a narrow, irregular site at the northeastern tip of the university campus (bordered by the university's football ground to the south, and the Chatham Road/Kowloon Corridor motorway interchange to the north), the JCIT is connected to the heart of the campus; encouraging the university's various faculties and schools to develop multidisciplinary initiatives and engagement with the community, government, industry, NGO's and academia.

The JCIT design dissolves the typical typology of the tower/podium into a more fluid composition. Interior and exterior courtyards create informal spaces to meet and interact, complementing the large exhibition forums, studios, theatre and recreational facilities. The tower's design promotes a multidisciplinary environment by connecting the variety of programs within the School of Design; establishing a collective research culture where many contributions and innovations can feed off each other.

Students, staff and visitors move through 15 levels of studios, workshops, labs, exhibition and event areas within the school. Interior glazing and voids bring transparency and connectivity, while circulation routes and communal spaces have been arranged to encourage interaction between the many learning clusters and design disciplines.

With its contribution of HK$249 million towards the construction of JCIT, The Hong Kong Jockey Club Charities Trust also funds the

Jockey Club Design Institute for Social Innovation.

Three dimensional drafting and Building Information Modelling (BIM) was used resolve the evolving geometry of Jockey Club Innovation Tower. Arup performed the role of structural and building services engineer to help realise the design by Zaha Hadid Architects. BIM was integral to the design and construction of this three dimensionally irregular-shaped building; enabling clash-analysis with building elements and services, providing an efficient mode of communication between the architectural, structural, building services and facade engineers. Arup used BIM to integrate the building services design with the building's unique internal geometries which could not be fully represented with conventional 2-dimensional drawings.

"The facade is one of the key elements of this building. Weekly design workshops were often focused on the verification of design information provided by the architect against BIM information built from the onsite surveying of the structural elements, building services and facade structures," an Arup spokesperson said. "BIM frequently highlighted design issues which would have been impossible for either the consultants or contractor to detect using traditional graphical projection drawings. With the assistance of BIM, the coordination was more efficient and avoided a lot of abortive site works."

Arup also used design software in Etabs to carry out the structural analysis. With the coordinates from the Etabs model, the engineers produced an outline of the structure to communicate with the Zaha Hadid Architects' 3-Dimensional computer model.

The structural engineering Innovation Tower included the tower overhanging the footpath on and the raked elevation north of the site, with foundations not permitted within the path. In response, Arup designed the superstructure to use three main cores and beam-column frames for lateral load and eccentric tower loads. Raking columns are used in some areas to handle loads from tilting. Discrete transfer beams have been used on the third floor to free up the lower two levels as much as possible.

The foundation of the tower used 54 large diameter bored piles, which minimised construction vibration and ground settlement that could affect adjacent buildings, and added capacity to resist very large loads arising from the tilting disposition of the tower. During construction, Arup established alarm monitoring levels during basement excavation in case of nearby settlement.

With a unique building geometry, Arup maintained Zaha Hadid Architects' design intent for the curved 3D facade by investigating several alternative cladding materials including fibre reinforced plastic, tensile fabric and aluminium. A unique three dimensional metal cladding was selected. A multiple layer of panelling allows the top layer

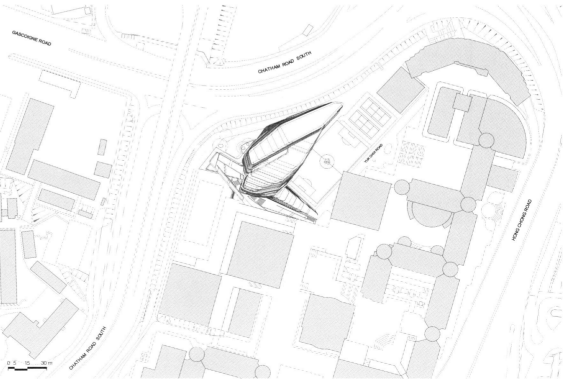

3 总平面

of cladding to have open joints, while a hidden lower layer provides functional weatherproofing and drainage functions.

The key facade design challenge on the project was the detailing and interface resolution of the curved 3D facade cladding. A series of catwalks were designed behind the facade cladding with removable access panels to the interior to access glass that could not be reached by the building maintenance unit's normal operation. The panels are located in spaces which can be safely closed by facilities management staff while the building is open.

Arup implemented differential pressure controller valves for the chilled water system for stable hydraulic performance and convenience for future alterative works. The chilled water circuit design allows the full and part load performance of chillers to be measured according to a process set out in a technical paper by Professor Francis Yik, from the university's Department of Building Services Engineering.

The building has also incorporated energy efficient features in building services installations, including evaporative cooling towers for water cooled chillers, energy wheels provided to the primary air unit to enable energy recovery between exhaust air and fresh air intake, and carbon dioxide sensors in general teaching facilities and offices to modulate the volume of fresh air supply.

Built-in thermostats sense average room air temperature from a sample of air induced into the unit. An energy humidity recovery ventilator is also included instead of a heat wheel to allow sensible and latent waste heat to be recovered. A central control and monitoring system manages all lighting and mechanical equipment to optimise energy saving.

4 通过弯曲的三维金属面板表皮材料实现了建筑独特的几何形态

4

5 二层平面

1 二层入口大厅
2 汽车设计实验室（TDL）
3 多媒体教室
4 模型工作室
5 办公室
6 室外景观

6 一层平面

1 一层入口大厅
2 木材车间
3 金属车间
4 摄影工作室
5 影视制作
6 汽车设计实验室（TDL）
7 地下通道

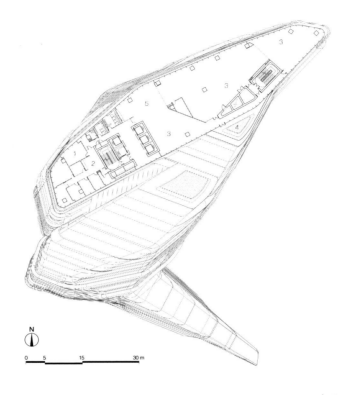

7

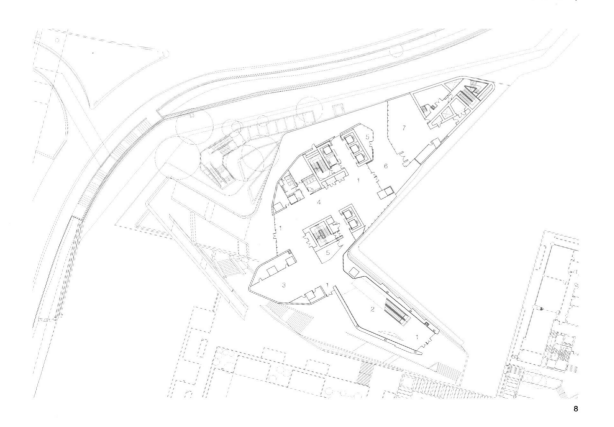

8

7 四层平面
1 会议室
2 办公室
3 展厅
4 屋顶平台
5 公共空间

8 三层平面
1 入口大厅
2 庭厅
3 画廊
4 临时展厅
5 商店
6 室外剧场
7 展览辅助用房

9 室内的玻璃隔断和中庭空间使各部分空间彼此通透

HIGH-RISE BUILDING | DESIGN WORKS

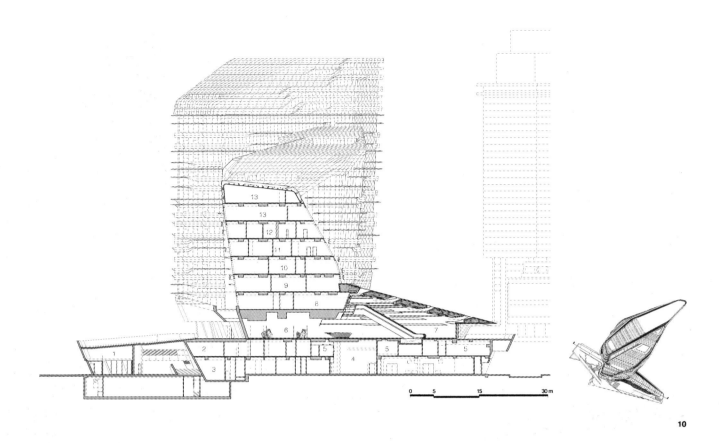

10

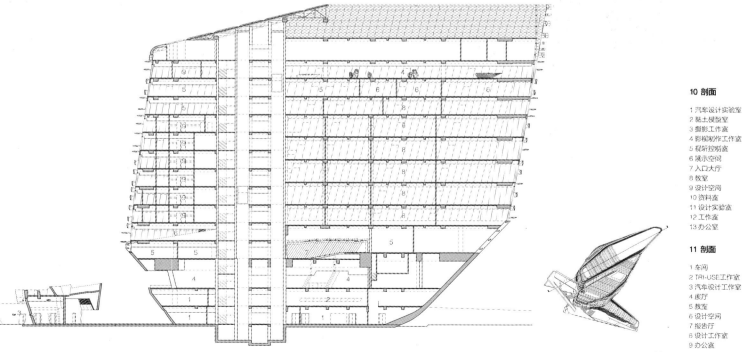

10 剖面

1 汽车设计实验室
2 黏土模型室
3 摄影工作室
4 影视制作工作室
5 视听控制室
6 展示空间
7 入口大厅
8 教室
9 设计空间
10 资料室
11 设计实验室
12 工作室
13 办公室

11 剖面

1 车间
2 TRI-USE工作室
3 汽车设计工作室
4 展厅
5 教室
6 设计空间
7 报告厅
8 设计工作室
9 办公室

11

12 楼内设有用于设计教学及创新的各种空间
13 精心设计的环形交通和公共空间激发了众多学习集群和设计学科间的交流

THE OPUS OFFICE BUILDINGS, UAE
阿拉伯联合酋长国 The Opus 办公大楼

扎哈·哈迪德建筑师事务所 | Zaha Hadid Architects

项目位置：Dubai, UAE
项目业主：Omniyat Properties
业主代表：Graham Hallett
建筑设计：Zaha Hadid and Patrik Schumacher
项目建筑师：Christos Passas
建筑负责：Vincent Nowak
项目团队：彭文苑，Chiara Ferrari，Javier Ernesto Lebie，Paul Peyrer-Heimstaett
项目经理：Gleeds（伦敦）

当地建筑师：Arex Consultans（迪拜）
机电工程：Whitbybird（伦敦）
防火设施：SAFE（伦敦）
立面工程：Permasteelisa
电梯顾问：Roger Preston Dynamics（伦敦）
交通流线：Cansult Limited（迪拜）
承包单位：Nasa / Multiplex
建筑面积：84 345 m²/100 m（W）× 67 m（D）× 93 m（H），21层

迪拜，是现今世界上除了中国之外的另一个建筑舞台，众多国际知名建筑师都在这个大舞台上占据一片天空，并向世界展现他们的建筑理念。扎哈·哈迪德在过去几年内开始挑战商业建筑，打破了过去30年来都是以文化类建筑设计为主的状态，并在迪拜拿下4个建筑设计项目，这使得她领导下的设计团队开始面临另一个全新的挑战——在以利益导向为优先的开发商业主面前，如何坚持对设计质量与设计理念的一贯追求。

2007年5月，扎哈·哈迪德在伦敦大英博物馆展出了她在迪拜的最新作品——The Opus。它是一栋高100 m、面积8.6万 m²、AAA等级的21层复合式办公大楼。Opus原是拉丁词，意指"作品"，大多指一件艺术作品，在建筑领域中它特指一种将建筑物内不同元素连接在一起的技术、方法或是风格。

设计概念

新的都市规划对现有迪拜的都市肌理提出一种以纪念性象征为主的风格，并提出在规划中要将一系列独特的建筑通过一连串建筑物的墩座和街道以及河道相连，创造出一个统一的、全新的区域——迪拜商业湾（Dubai Business Bay）。因此本案在整体规划上寻找一种既有连续性又保有各自独特性的提案，由此提出一个具备独特、多元和流动特质的空间概念，其可以向外延伸并与周围环境中一系列重要建筑形成独特的联系和互动。

整体规划

本基地是迪拜开发集团——Omniyat的旗舰级办公大楼，两块基地由低楼层的基座结构体连接成一体，并将原属于两个独立的体块诠释为一栋完整的建筑。将两块基地视为一体，并将两栋建筑的功能合并，企图使建筑体的边界被阅读性模糊化，而建筑物地面层四周有着缩进的廊道，相互串连并延续着地面层行人的动线。

二元性

单一的形体存在着二元的特质，分别表现在机能、形体和构造中。单一的立方体、单一的曲面空间，在建筑体中是两种截然不同的本质。设计师透过中间曲面围塑成的虚空间将两个体量整合在一个完整的立方体中，而中间被挖去的曲面空间，是立方体中的虚体，也可被视为视觉上的实体。

动与静、虚与实、刚与柔、内与外、简单与复杂、冲突与共生、个性与共性，在这雕塑般单一的建筑中通过形体和材质的处理找到一种动态的平衡。

立方体

硬朗的轮廓创造出建筑的实体感，清楚界定建筑与都市的边界；静止的个体由虚的透明玻璃构成，整合并分享着原本三个独立的体量。

曲面虚空间

由黑色玻璃构成的柔软形体具有自由的轮廓，其独特的几何形体在立方体中形成一种独特的能量，流动、切割在曲面的虚与立方体的实之间创造出一种动态的空间感。曲面流动的轮廓勾画出不同的都市景观，开始了建筑与都市之间动态的对话。

地面层

缩进的地面层空间和自地面延伸并跃起的斜坡，划开了立方体与地面的联系，让其静止的主体漂浮在流动的地面上。透明的开放地面层空间由不同动线构成了复合的内在层次。动线被汇集到内部挑高 20 m 的中庭空间后，再通过上升的大斜坡分别向两侧的独立大厅移动，将办公和商业行为的动线分隔开来，同时保持着视线上的连续性。

立面

立方体的透明玻璃上有着镜面反射效果的图腾，在立面上形成丰富的层次，不仅提供特定程度的反射性和物质性，同时具有减少室内空间热度的功用。

半透明的镜面反射玻璃在建筑表皮上创造出不同的透明度，让立方体通过反射的方式融入周围环境中。模糊个性与共性的关系，让建筑同外围环境有着另一个层次与深度的对话，并进一步凸显出曲面虚体的实体性。

灯光

通过灯光的营造，这栋大楼给人感知上的交替感觉：白天，立方体是实的，而曲面虚体是空的；而到了夜晚则完全相反，立方体变得很暗并且去物质化，而曲面虚体却被灯光活跃起来，即便在很远的距离下依旧清晰可见。相信通过设计上的巧妙安排，The Opus 将赋予迪拜一个独特、多变和流动的新空间和新地标。

正如扎哈本人所言："The Opus 是一栋具有流动性、空间性的建筑，是一栋拒绝接受传统定义下办公空间机能性建筑……我们应用最先进的科技和最独到的设计策略，让使用者体验到一个更优质的工作环境……"

1 入口广场

2 临水立面
3 各角度建筑效果

高层建筑 | 设计作品

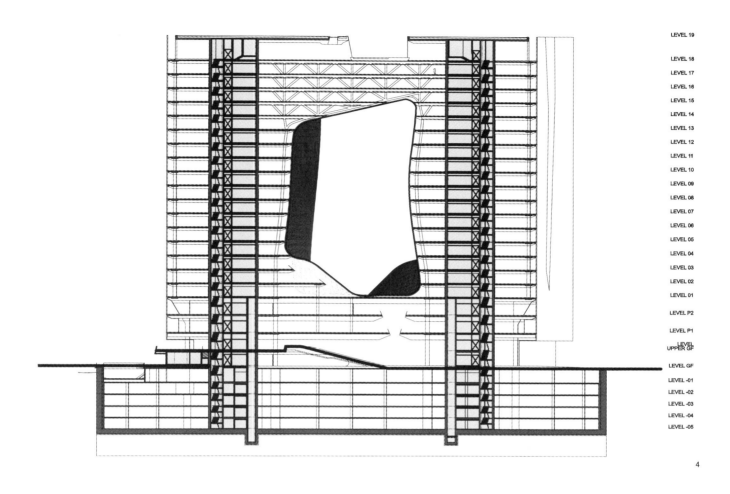

4

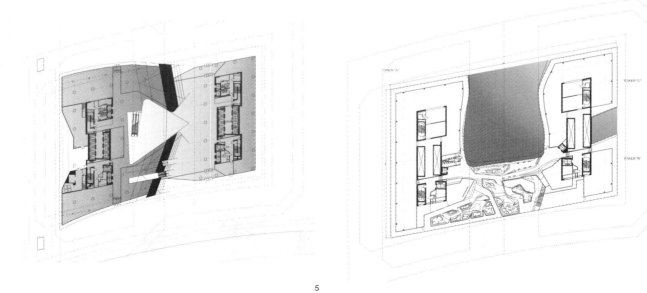

4 剖面图
5 首层平面
6 七层平面

5　　　　　　　　　　　　6

109

# SUNRISE TOWER, MALAYSIA
马来西亚黎明之塔

扎哈·哈迪德建筑师事务所 | Zaha Hadid Architects

项目客户：Sunrise Berhad
项目建筑师：Zaha Hadid Architects
项目设计：Zaha Hadid with Patrik Schumacher
项目指导：Tiago Correia
项目经理：Victor Orive, Fabiano Continanza
设计团队：Alejandro Diaz, Rafael Gonzalez, Monica Noguero, Oihane Santiuste, Maren Klasing, Martin Krcha, Daniel Domingo
机械工程：Buro Happold
当地建筑师：Veritas Design Group
QS/项目估算：JUBM
基地面积：6 800 m²
建筑面积：150 000 m²

由扎哈·哈迪德建筑师事务所设计的黎明之塔与城市环境实现了多元化的功能对接。通过对不同层面潜在协同作用的不断探索，建筑设计立足现有城市环境，创造了一个与周边环境及建筑接轨的服务平台。设计方案将全部功能融于一座建筑之中，通过运用细致的景观策略，实现了建筑与地面的过渡，连接了基址的各个部分，完成了新人行道路与内街路体系的整合，从而构建了全新的建筑结构。

建筑的功能包括公寓住宅、酒店、写字间、零售商业区以及停车场。这些区块间的衔接是项目的核心问题，我们借此呈现出一个构思精致、规模与特色兼具的建筑设计方案。各项功能或采取分层堆叠方式层层展开，或根据建筑的平行分支分配。各功能的协同是通过将特定功能区块加以融合实现的，这一方法普遍应用于大型空间的设计建造，它们不仅对公共与私有进行了区分，而且通过必要安全措施的引入实现了环境之间的无缝转换。

由核心设计理念所营造的建筑空间融汇了楼体和景观，强化了室内与室外环境的流畅过渡。建筑主体的外立面表现力突出，满足了功能、结构以及楼体外墙的各项要求。参数设计阶段引入了空间网的设计手段，使得建筑各个部分充分做到了本地化。各个组成部分的详细规划设计工作通过综合地形、朝向、功能以及结构负载等多项因素生成的楼面网得以完成，产生了从基座突出的对角线结构向顶层和缓的垂线渐变的独特视觉效果。

功能参数设计同样围绕居住空间网以及公寓建模系统进行规划。根基空间网设计在建筑平面布局和结构中起核心作用，各区块可相互转换，从而根据委托方要求构建出灵活多变的景观样式，具备适应长、短期各类变化的能力。各部分结构框架支撑楼体的外立面，同时勾勒出了整座建筑优雅大方的造型。设计综合了自然采光、遮光、功能、通道入口及观景视线等诸多因素，通过建筑外观这一关键元素实现了室内外空间的过渡。

设计方案借助一系列独立的流程图规划管理着整座大厦内各个功能区块间的不同路线。通过这些路线的连接，过道以及公共设施层起到了交通枢纽的作用，它们如同十字路口一般，令功能和位置的转换自由方便。同立面的设计类似，循环体系的创建也是通过多维空间网实现的，设计包含了功能转换和实现室内外空间的无缝对接，使得整个方案清晰明确，各层面衔接自然。起着楼内走廊寻向标志作用的天井及公共区域设计不但建立了核心功能区块间的视觉联系，同时还保证了所有人员均有机会使用楼内的各项设施。

大厦的各个功能区块分置于66层楼面之内，其中地下4层，地上62层，绝对高度达280 m。地面大厅是整座建筑的核心枢纽，经规划成为了公寓、酒店、写字楼以及公共区这4个不同的功能区。

1 临街效果

Zaha Hadid Architects' design for Sunrise Tower engages with the city in multiple ways. By exploring potential synergies at different levels and anchoring itself to the existing urban fabric, it creates a platform of services that engage with neighbouring developments, sustaining critical mass and a sense of community. The scheme merges all programmes into one building, distancing itself from the traditional tower and podium typology. Through a detailed landscape strategy the design interweaves tower and ground, extending and connecting the different parts of the site, integrating the new pedestrian routes and internal road system, structuring the fabric of the new development.

The design houses 5 different programmatic components: residential, hotel, offices, retail and parking. Connectivity between these parts becomes central to the project in order to produce an articulated design that encompasses both the scale and the different qualities of each of the parts, fusing them into a coherent scheme. The program is stratified, stacking one function over the other, or carrying them in parallel when the tower branches. Programmatic synergies are created by blending certain programmatic aspects that are common to create powerful spaces that not only differentiate between programs but also enable a system that separates public from private, yet integrating the necessary security features to organize a seamless transition between environments.

The design concept creates spaces that blur the difference between building and landscape, intensifying the fluidity between interior and exterior. The tower body is developed through a performative outer skin that merges programmatic, structural and building envelope requirements. A spatial grid is generated through parametric component design, enabling the local adaptation of each component to accommodate for different requirements within a pool of repeated elements. Components variation is choreographed through the floor grid with a rhythm that is defined by topology, orientation, programme and structural load, generating a customized gradient that mutates from strong

2 临街效果
3 形体细部

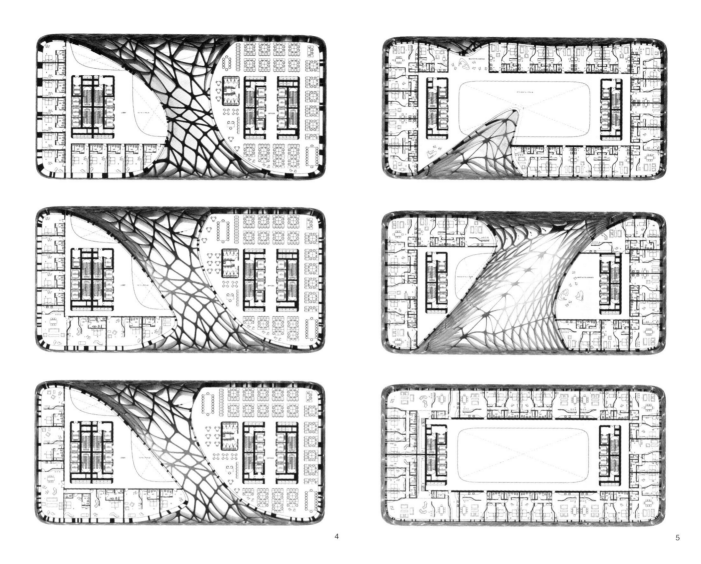

4 十二、十九、二十一层平面/酒店公寓及办公空间
5 三十三、四十六、六十二层平面/住宅

diagonals at the base to gentle verticals at the top.

The programmatic parameter also develops a grid of occupation, a modular system for hotel rooms and residential apartments. Because this grid is integral to the buildings topology and structure, tube-in-tube system, the partitioning system becomes interchangeable, constructing a flexible landscape to be tailored to the client's requirements, capable of assimilating changes in short, medium and long terms. The created pattern is supported through the structural frame elements, carrying the building's facade and generating interesting elegant views through the tower. Its design concept integrates natural light, shading, program, access and views, making the component the key operator of the transition between interior and exterior spaces.

The building is designed through a series of independent flows that map the tower and organize different routes for different programmes. Along these routes the lobby and shared facilities floors work as communication hubs, like intersections that enable flexible itineraries and changes between uses. Similarly to the skin, the circulation materializes as a multidimensional spatial grid, inclusive of the program, treating interior and exterior in a seamless way, thus maximizing the clarity of the scheme and the perception of the different levels. The design of a clear navigation system for lobbies, atria and common areas, enables visual communication as well as access through the cores, ensuring fully accessible environment for all users.

The building's complex programme is distributed through 66 floors in total, 4 bellow ground and 62 above ground, with an absolute height of 280m. The ground lobby is the primary hub of the tower, defining 4 different dedicated lobbies for residential, hotel, offices and general public.

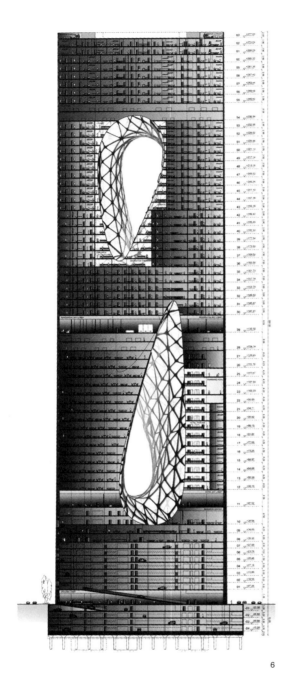
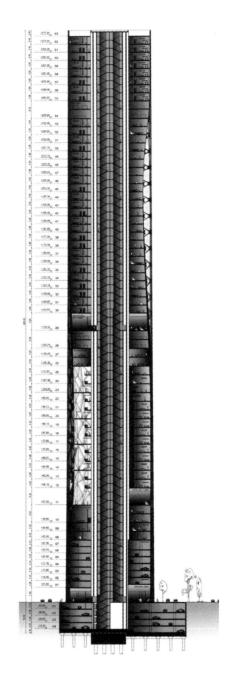
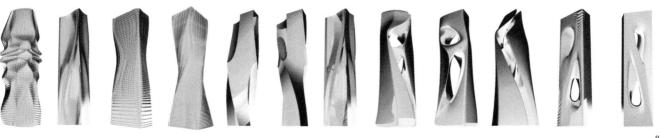

6　剖面
7　剖面
8　形体演变

9 竖向空间效果
10 休闲空间效果
11 入口大厅效果
12 交通空间效果

DOROBANTI TOWER, ROMANIA
罗马尼亚多罗班蒂大厦

扎哈·哈迪德建筑师事务所 | Zaha Hadid Architects

建筑设计：Zaha Hadid with Patrik Schumacher
项目建筑师：Markus Planteu
项目团队：Dennis Brezina，Naomi Fritz，Susanne Lettau，Thomas Mathoy，Goswin Rothenthal，Rooshad Shroff，Seda Zirek
项目功能：五星级酒店和公寓

多罗班蒂大厦的设计目标是力图在布加勒斯特中心地段建设一座地标性建筑。这座全新的超高层建筑融合了与众不同的造型设计、匠心独具的建筑结构以及高空生活的品质特色，其纯粹的金刚石削角造型必将成为布加勒斯特市中心精致高雅、历久弥新的地标。

扎哈·哈迪德建筑师事务所的设计理念是集建筑设计与结构工程为一身，结合独一无二的迂回网孔架构，自然流畅地表达酒店、配套设施以及公寓住宅等功能的变换。

建筑基址地处布加勒斯特市中心，东临罗马纳广场（Piaza Romana），北距国际机场约6 km。方案计划在多罗班蒂大酒店（Calea Dorobanti）与圣米海尔·艾米内斯库大教堂（St. Mihail Eminescu）接合部开发建设一处10万 m²的多功能建筑，包括一座3.4万 m²的五星级酒店（内设餐厅及会议中心）和总面积3.5万 m²的豪华公寓住宅。除此之外，设计方案还提供了4 600 m²的低层零售商业区，同时划分出大面积的公共空间，这些公共空间将是布加勒斯特独一无二的，这座10万 m²的综合型开发项目地处布加勒斯特市中心，集地下停车场、赌场、零售商业区、餐厅、会议中心、酒吧、文娱设施以及公寓住宅等功能于一身，展现着布加勒斯特的城市魅力。

建筑设计

城市环境、基址限制以及建筑规划等因素共同造就了建筑精致优雅的锥形轮廓。楼体独特的几何造型反映了布加勒斯特的城市结构，与其过去政治体制下的畸形发展形成了鲜明的对照。全新的超高层建筑在保证设计与众不同的同时，避免了外观的重复。建筑整体呈削角金刚石梭形，体量中部半径最大，逐渐向体量顶、底两端回缩变细。楼体顶部的空隙保证了室内充足的自然光照，周边环境的景观视线不受阻挡，而底座的壁阶则提供了充足的公共空间及建筑前部的专用入口。

根据要求，毗邻多罗班蒂大酒店的两座豪华住宅必须保留；方案规定基址四周界线需后退3~5 m；工程委托方要求保留建筑东北角现有建筑。以上种种限制最终促成了建筑狭窄的齿状造型设计。

建筑立面与结构设计

建筑自上而下采用弯曲波浪式钢筋混凝土结构，形式别具一格，成为多罗班蒂大酒店有效的衬托和补充。混凝土填充技术提升了整体架构的强度以及钢结构的耐火能力。立面结构也根据功能的区分进行了相应调整。建筑基座部分的立面网状结构根据建筑高度的结构性需要增加了密度，达到了规划的承重及硬度要求。

除此之外，次级结构也为主钢结构提供了支撑，而且次级结构还因为与各层楼面高度一致而成为这座200 m高的超高层建筑的人体比例参照物。另外，还可利用其作为附加玻璃镶板的支架，一并组装成为遮光装置。

The Dorobanti Tower was designed to establish an iconic presence in the heart of Bucharest. The new tower is a unique mix of a distinctive form, ingenious structure, and spatial qualities of sky-high living. The purity of its form - a chamfered diamond like structure - will be a timeless, elegant landmark in the centre of Bucharest.

Zaha Hadid Architect's design concept is a synthesis of architecture and engineering, which integrates a distinct meandering structural mesh frame and naturally expresses the changing programme of hotel, amenities, and residential apartments.

The site is located in the centre of Bucharest, to the west of Piaza Romana, and approximately 6km south of the international airport. The brief called for a 100,000 square metre mixed-use development at the junction of Calea Dorobanti and St. Mihail Eminescu. The project comprises 34,000 square metres of a 5-star hotel (including restaurants and a convention centre) and 35,000 square metres of luxury apartments. Additionally, the scheme offers lower level retail areas of 4,600 square metres and it delivers a generous allocation of public realm. This public area will be unlike anything else in Bucharest, representing a major attraction within the dense urban character of the City, offering an important new meeting space and urban plaza. The 100,000m^2 mixed-use development in the centre of Bucharest comprises underground parking, a casino, retail, restaurant, convention centre, bar, recreation facilities, and residential apartments.

Urban parameters, site constraints and the building programme generate the building's elegant tapering profile. The unique building geometry responds to the urban structure of the city and creates a counterpart to the angular developments of the communist past of Bucharest. The new tower establishes a distinctive identity while avoiding sterile repetition through its dynamically changing appearance. The chamfered diamond shape tapers from the centre towards the top and the bottom. On top of the structure, the recess assures more sunlight and views for the surrounding neighbourhoods, while the offset at ground level creates public realm and an appropriate entrance plaza in front of the tower.

It is required that the two listed villas adjacent to Calea Dorobanti shall remain in place. Due to planning regulations, the site perimeter

1 建筑全景

2 夜色中通透的体量效果

line is offset between three and five metres. Furthermore, the client requests the northeast corner to remain undeveloped with buildings. All the above restrictions lead to a narrow and jagged building plot.

Concrete filled steel profiles follow in sinus waves from the ground level to the top of the tower, creating a distinctive identity and complementing the tower design. The concrete filling will give additional strength to the structure and it will provide fire protection to the steel profiles. The facade structure adjusts to the building programme and to the structural forces. At the bottom, the facade grid has denser amplitudes according to the structural requirements for a tower of this height, providing the required load bearing capacity and stiffness to the structure. At the technical and recreation levels, the structure condenses creating almost solid knots.

Additionally, the secondary structure supports the main steel frames. It also gives the 200m tower a human scale as the grid of the secondary frame structure reflects the floor heights. Furthermore, the secondary structure could be utilized to support additional glass panels as a shading device.

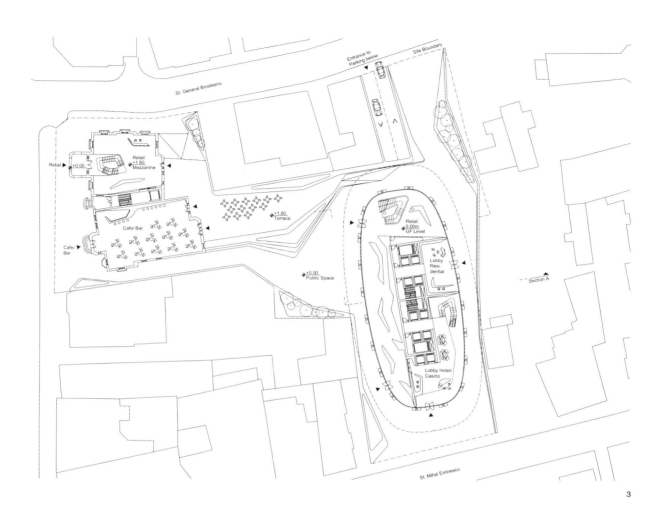

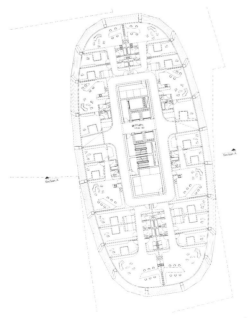

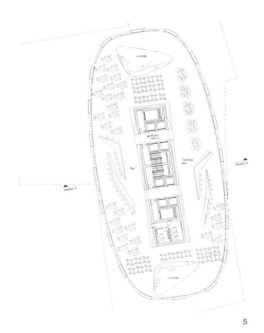

3　首层平面
4　公寓标准层平面
5　二十九层平面

HIGH-RISE BUILDING | DESIGN WORKS

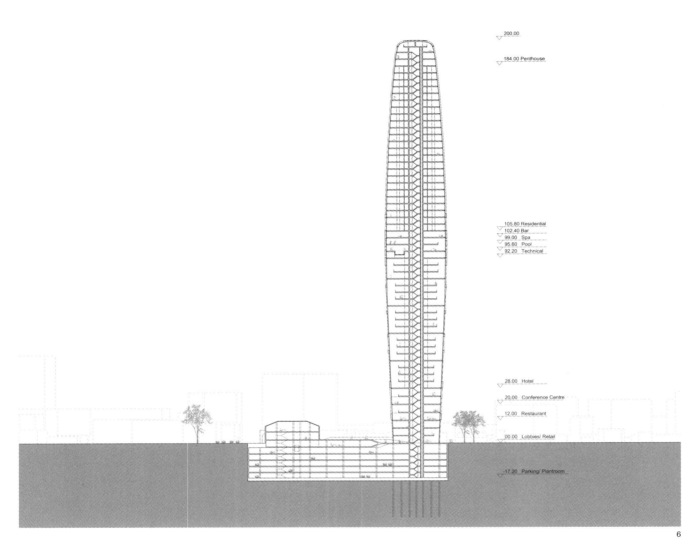

200.00
184.00 Penthouse
105.80 Residential
102.40 Bar
99.00 Spa
95.60 Pool
92.20 Technical
28.00 Hotel
20.00 Conference Centre
12.00 Restaurant
00.00 Lobbies/ Retail
-17.20 Parking/ Plantroom

6

7

8

6　剖面
7　独特的结构表皮
8　弯曲波浪式的结构
9　休闲空间
10　公寓内景
11　入口大厅

9

10

11

HIGH-RISE BUILDING | DESIGN WORKS

NILE TOWER, EGYPT
埃及尼罗塔

扎哈·哈迪德建筑师事务所 | Zaha Hadid Architects

建筑设计：Zaha Hadid with Patrik Schumacher
项目建筑师：Joris Pauwels
设计团队：Feng Xu、Paulo Flores、Sharifah Alshalfan、Tariq Khayyat

全新的尼罗塔体量设计轻盈而优雅，建筑基址地处开罗市中心，毗邻尼罗河。

这座70层的酒店及公寓建筑主结构采用混凝土翼缘墙体，墙体随建筑高度升高而徐徐旋转，令酒店客房以及公寓住宅具备多种朝向，可供入住者选择，同时令建筑外观造型动感十足。

从建筑西侧面可俯瞰尼罗河，因此尼罗塔高层楼面向外突出，为西侧客房及公寓提供了最广阔的景观视野。东侧低层楼面向外拉伸，呈三角形，以保证建筑体量的最大稳定性。这一相对的外延结构令整座建筑的外观好似向尼罗河倾斜一般。

建筑的几何造型在强化外观的同时，也给驻足于楼宇之下、翘首高处的人们以尼罗塔仿佛在运动的感官体验。而这一徐徐转动的外观又通过南北立面突出的对角装饰线得到了进一步的强化。西立面造型围绕低层楼体展开，整体悬于滨水街之上，容纳了尼罗塔的各项公共功能单元。立面结构在到达基座位置时，将除楼体垂直占地以外的其他剩余空间一并囊括其中，延伸形成了一座玻璃广场。尼罗塔不但为尼罗河东岸增色添彩，还为开罗的天际线添上了宝贵的一笔。

The new Nile Tower is designed as a gracious volume that elegantly borders the Nile River in the heart of Cairo.

The main structural elements of this 70 storey hotel and apartment tower are concrete fin walls that rotate gently over the full height of the tower. The gradual rotation of the fin walls generates a wide selection of orientations for the hotel rooms and apartments. These fin walls are reflected on the external facade, giving the tower its dynamic appearance.

On the west side – overlooking the river - the tower bulges out at the upper floors to maximize the views from hotel rooms and apartments. On the east side the tower is pulled out as a triangle on the lower floors to offer maximum stability. These opposite extensions make the overall tower seemingly lean over towards the Nile.

While enhancing the perspective when looking up from the foot of the tower this geometry also makes the tower seem in motion. This seeming rotation is further enhanced by emphasizing the diagonals running along the north and south facades. The facade geometry on the west facade culminates at the lower levels in a volume overhanging the road towards the river, housing the common program of the tower. At the foot of the tower, the facade drapes down over the large column free space next to the tower footprint where it forms a continuous glass plaza. A high quality contribution to the east bank of the Nile, The Nile Tower also forms a valuable addition to the skyline of Cairo.

1 夜景效果

3

4

2 滨水效果
3 入口空间
4 充满动感的内部空间
5 交通空间

5

HIGH-RISE BUILDING | DESIGN WORKS

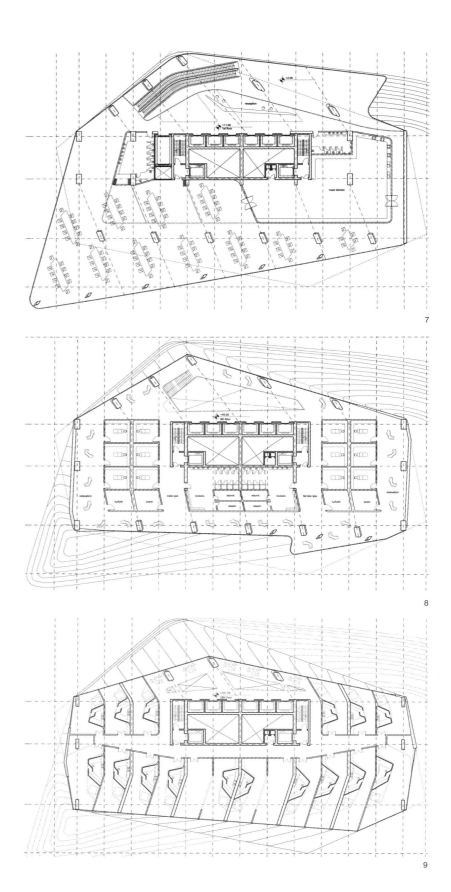

| 126

6 临街效果
7 一层平面
8 八层平面
9 二十五层平面
10 剖面图
11 剖面图
12 总平面

FARRER COURT, SINGAPORE
新加坡花拉阁

扎哈·哈迪德建筑师事务所 | Zaha Hadid Architects

项目客户：新加坡 CapitaLand
项目功能：居住建筑和景观设计
项目规模：地上面积 22 万 m²，地下建筑面积 7 万 m²
建设高度：150 m
项目设计：Zaha Hadid Architects
建筑设计：Zaha Hadid with Patrik Schumacher
项目负责：Michele Pasca di Magliano, Viviana Muscettola
项目经理：Charles Walker
设计团队：Edward Calver, Bianca Cheung, Dominiki Dadatsi, Loreto Flores, Helen Lee, Hee Seung Lee, Jeonghoon Lee, Feng Lin, Jee Seon Lim, Ludovico Lombardi, Clara Martins, Kutbuddin Nadiadi, Hoda Nobakhti, Annarita Papeschi, Line Rahbek, Hala Sheikh, Zhong Tian, evil Yazici, Effi e Kuan, Sandra Riess, Eleni Pavlidou, Federico Dunkelberg, Evan Erlebacher, Gorka Blas, Bozana Komljenovic, Sophie Le Bienvenu, Kelly Lee, Jose M. Monfa, Selahattin Tuysuz, Ta-Kang Hsu, Emily Chang, Judith Wahle

当地建筑师：RSP, Singapore
结构工程：Maunsell, Singapore
M&E 工程：Max Fordham /Londo（概念设计），M&E Engineering BECA Singapore, Quantity Surveyor DLS/Singapore
景观设计：GROSSMAX, Edinburgh（概念设计）
景观建筑师：ICN/Singapore
机电照明：LPA/Tokyo
声学设计：Engineering Acviron/Singapore

建筑基址

花拉阁基址地处新加坡住宅区的关键地段，毗邻荷兰路（Holland Road）及未来规划中的MRT车站，与交通干道花拉路（Farrer Rd.）直接连通，基址周边无高层建筑。如此优越的地理及交通条件使该项目广受全城民众的关注。

规划设计

扎哈·哈迪德建筑师事务所经过对现有基址的平面结构和四周主要轴线的研究，最终确定了花拉阁规划方案。方案根据主要轴线设计了一系列与基址周边环境深度融合的网格。地面景观以绿色为基调，旨在凸显新加坡独特气候所孕育的丰富植被资源。基址各层按照阶梯式高地形式进行重新组织规划，最大限度地拓展了公共配套设施用地。建筑的方位及朝向也根据当地气候条件进行了优化调整，同时保证了周边城市景观不受遮挡。

建筑设计

设计方案中的7座超高层建筑自基址景观中的洼地私人园林之中拔地而起。各低层楼面扭结相连，着力突出楼体同地面的连接，从而创造出更大面积的公共空间，并根据整个项目的规模和建筑密度营造出独具特色的私人景观园林。本着建设多座形态各异、独一无二的高层建筑的基本原则，各楼体根据每层住宅单元的数量分成了若干个垂直面，并采用立体切割手法，令建筑造型极具三维立体感。与此同时，这一设计最大程度地解决了楼面空间的通风问题。建筑顶层运用高低不一的横梁组合完成了建筑顶端与天空背景的过渡。设计方案通过对基址内各楼体旋转角度的调整，以及阳台和立面镶板的细致推敲，在营造形态统一的高层建筑的同时，将不计其数的多样化设计手法成功地融入到了整个项目之中。

The Farrer Court site is located in a strategic position within the residential area of Singapore, close to the amenities of Holland Road and the future MRT station. The absence of high rise buildings in the near surroundings and direct connection to the main traffic route of Farrer Road make this a prestigious and highly visible site across the whole city.

ZHA's Proposal for the Farrer Court site is generated by the study of the existing alignments and the main axis surrounding the site, which are brought in and connected to generate a series of construction lines highly connected to the neighbourhood. The ground landscape level is visualized as a very green layer, which wants to emphasize the presence of florid vegetation in the Singapore's climate. The site levels are re-organized into a series of terraced plateaus to maximise the area dedicated to communal site amenities. The orientation and placement of the buildings is optimized in relation to the local environment as well as to maximize views out towards the surrounding city and landscape.

The program is organized into 7 towers, which grow from sunken private gardens within the site landscape. The lower floors kink in to highlight the point where buildings meet the ground, enabling yet a greater open area and the creation of highly private gardens which are quite unique given the scale and density of the development. The towers are subdivided into petals according to the number of residential units per floor, with a common principle a series of diverse and unique towers can be generated. The petals are expressed in three dimensions thanks to vertical cuts which give definition to the building's facades and, at the same time, allow for cross ventilation of most of the flats. The buildings culminate at the top with a series of fingers stepped at different heights, which blend the transition between the architectural fabric and the sky. Through rotating the buildings across the site, and the careful use of balconies and facade panelling a combination of self similar towers produce an incredible amount of diversity across the development.

1 花拉阁全景

2 临街效果

3 建筑与景观的有机融合
4 总平面
5 塔楼 A 平面
6 塔楼 B 平面

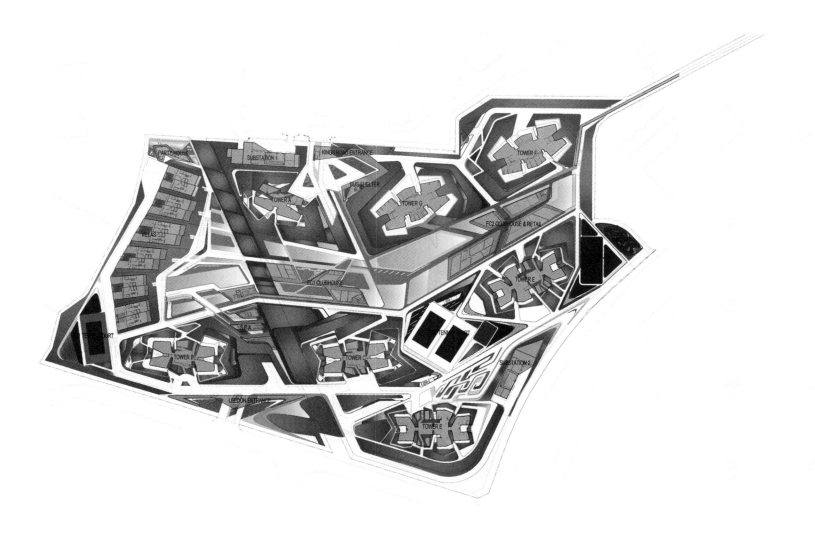

4

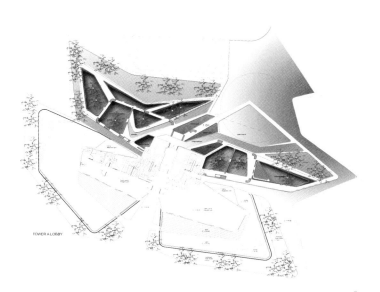

5

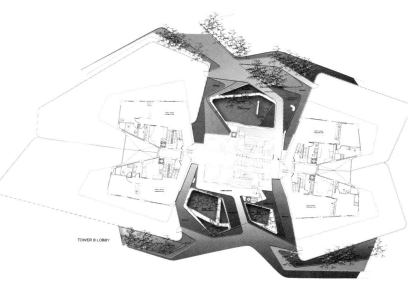

6

7 局部景观
8 局部景观

高层建筑 | 设计作品

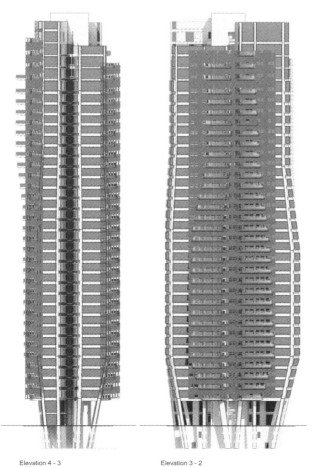

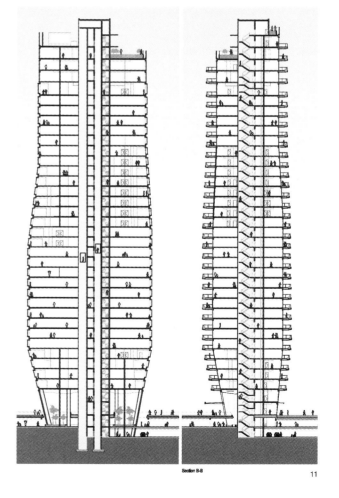

9　花拉阁夜景
10　塔楼 A 立面
11　塔楼 A 剖面

GUANGZHOU INTERNATIONAL FINANCE CENTER
广州国际金融中心

威尔金森·艾尔建筑事务所 | Wilkinson Eyre Architects
项目获2012年英国皇家建筑师学会（RIBA）国际建筑莱伯金奖（Lubetkin Prize）

项目名称：广州国际金融中心
业　　主：越秀地产股份有限公司
建设地点：广州市天河区珠江新城珠江西路
建筑面积：44.8万 m^2（其中塔楼建筑面积24.7万 m^2，办公面积16.8万 m^2）
建筑层数：103层
建筑高度：440 m
项目投资：2.8亿英镑
设计单位：Wilkinson Eyre Architects

设计总负责：Chris Wilkinson, Dominic Bettison
结构工程：Arup
设备设计：Arup
设计时间：2006年1月
建成时间：2010年10月
图纸版权：Wilkinson Eyre Architects
摄　　影：Jonathan Leijonhufvud, Christian Richters, Will Pryce

　　2004年威尔金森·艾尔建筑事务所在国际竞赛胜出，负责设计440 m高的广州国际金融中心。建筑外形简洁、优雅，作为广州珠江新城主轴的标志，连接起北部的商业区和南部的珠江。三角形平面使建筑获得面向珠江的最大景观视野，且满足高效的内部空间布局和卓越的环保性能的需要。塔楼共103层，功能空间丰富，由下至上分别为办公区、豪华酒店和顶层观光区。主厅通过自动扶梯与位于地下层的辅助办公大厅相连，从而为前往地下的商业区和地铁站提供了通道。在地面层，塔楼与一个包含了零售商场、会议中心和具有高端公寓的综合体裙楼相连。

　　塔楼所采用的全球最高的斜肋构架结构通过立面清晰地表达出来，赋予建筑鲜明的特点。斜肋构架的构件由混凝土填充的钢管制成，具有良好的结构刚度及防火性能。管状斜肋结构每12层形成一个节点，从而构成若干54 m高的巨大的钢铁钻石。在塔楼底部，结构构件的直径为1 800 mm，随着高度的增加尺寸逐渐减小，达至顶层时构件直径只有900 mm。

　　结构核心承担了楼板的绝大部分重力荷载，并通过楼板梁与斜肋构架外围结构相连，形成一个坚固的"筒中筒"结构系统。结构的固有刚度实现钢材用量最小的同时还能抵抗加速和摇摆的影响，从而为建筑使用者提供了较高的舒适度。这种刚度和对加速的抵抗意味着建筑需要一种无阻尼的结构。塔楼的外形设计能够减弱风力影响，从而减少了必要的结构尺寸和重量。

　　塔楼顶部的30层为广州四季酒店，344间豪华客房和套房围绕高度超越伦敦圣保罗大教堂和纽约自由女神像的壮观通高中庭而设，作为当地最现代、最宽敞的客房，在其中可尽享珠江美景。

　　落地窗不仅令室内光线充足，更为酒店的时尚餐厅、酒吧和客房提供了非凡的全景视野。具有雕塑感的中庭令人联想到塔楼钻石状的斜肋钢构架结构。在夜间，未来主义风格的照明设计艺术化地点亮了整个中庭的所有围栏，更强调了结构的特点。

　　客人从位于1层的酒店大堂乘坐专用快速电梯到达位于70层的空中大堂，一刹那便会迷醉于珠江畔的美景和高处天窗倾洒而下的阳光。位于第100层的餐厅和第99层的酒吧及高级俱乐部带给宾客无比强烈的高度感，两层之间通过悬挑在中庭的一段楼梯相连。

　　酒店还提供设施齐全的水疗中心、健身中心以及可以饱览壮观城市景色的巨大泳池。3个正式的宴会厅可为婚礼、社交活动和会议提供超过3 500 m^2的活动空间。

　　威尔金森·艾尔建筑事务所创始主管Chris Wilkinson说道："在广州国际金融中心的创作过程中，我们希望设计一栋简洁优雅的建筑，以丰富城市的天际线。这栋建筑与其他许多摩天大楼不同，它拥有一个光滑的具有空气动力学特性的玻璃外墙。而其引人之处源自于巨大的尺度和斜肋结构清晰、直观的表达。"

Wilkinson Eyre was selected in 2004 to design the 440m Guangzhou International Finance Center following an international design competition. Conceived as a simple, elegant form the tower marks Guangzhou Zhujiang New Town's main axis, which links the commercial district in the north with the Pearl River to the south. Its triangular plan maximises views of the Pearl River and responds to the need for efficient internal space layouts and excellent environmental performance. With 103 storeys, the tower has a mixture of uses including office space, a luxury hotel and a top floor sightseeing area. Within the tower, office floors occupy the levels below the hotel. The main lobby connects via escalators to a secondary office lobby located at the lower basement level, which in turn allows access to below ground retail and the subway station. At ground level, the tower connects with a substantial podium complex containing a retail mall, conference centre and high quality serviced apartments.

The building utilises the world's tallest constructed diagrid structure which is clearly expressed through the building's facade and gives the building considerable character. The diagrid members are formed from concrete filled steel tubes which provide both good stiffness and fire protection to the structure. The tubular diagrid structure "nodes-out" every 12 storeys to form 54m high giant steel diamonds. At the base of the tower the structural members are 1800mm in diameter and reduce in size up the building to 900mm at the top of the building.

The structural core takes much of the gravity load of the building's floors and is linked back to the diagrid perimeter structure via floor beams to create a stiff "tube-within-tube" structural system. The inherent stiffness in the structure minimises steel tonnage whilst providing resistance to acceleration and sway, thereby maintaining high comfort levels for the building's occupants. This stiffness and resistance to acceleration means that no damping of the structure is required. The shape

1 夜色中的金融中心

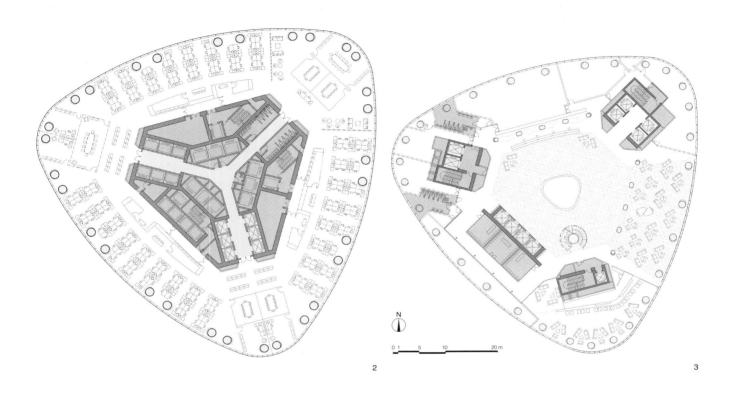

2 二十九层平面
3 七十层平面

of the building has been designed to reduce the effects of wind, thereby reducing the necessary size and weight of the structure.

Rising 103 storeys above the Pearl River, the Four Seasons Guangzhou occupies the top thirty of the new Guangzhou IFC. Arranged around a breathtaking full-height atrium, itself taller than St Paul's Cathedral in London or the Statue of Liberty in New York, the 344 luxurious guest rooms and suites are among the most modern and spacious in the city, with unrivalled views of the Pearl River.

The building's floor to ceiling windows make for light-filled spaces and create extraordinary panoramic views for the hotel's stylish restaurants, bars and guest rooms. The atrium balconies have been sculpted by Wilkinson Eyre so that when viewed from the atrium they recall the building's diamond shaped steel "diagrid" structure. This is accented at night by futuristic lighting design which artistically illuminates the handrails throughout the atrium.

Guests access the hotel's ground floor lobby and enter dedicated express elevators to the 70th floor sky lobby which benefits from amazing views out over the Pearl River and a towering atrium. This space is flooded with daylight from a dramatic roof light floating 120 metres above hotel guests' heads. Nowhere is the sense of height more intense than from the 100th floor restaurant and 99th floor bar and Executive Club lounge, with a staircase that is cantilevered over the atrium connecting the two levels.

The hotel offers a fully equipped spa, fitness centre and an infinity pool with spectacular views of the city below. Three formal ballrooms give over 3,500 sq meters of event space for weddings, social occasions and conferences.

Chris Wilkinson, Founding Director, Wilkinson Eyre Architects said: "In designing the Guangzhou IFC we aimed for an elegant simplicity that complements the City's skyline. It is quite different to many super high rises in that has a smooth aerodynamic glazed skin. The drama comes from the sheer size of the building and the clear expression and visibility of the diagrid structure."

4 塔楼所采用的全球最高的斜肋构架结构通过立面清晰地表达出来,赋予建筑鲜明的特点

5 位于第70层的空中大堂
6 空中大堂高处天窗倾洒而下的阳光
7 斜肋构架的构件由混凝土填充的钢管制成，具有良好的结构刚度及防火性能

高层建筑 | 设计作品

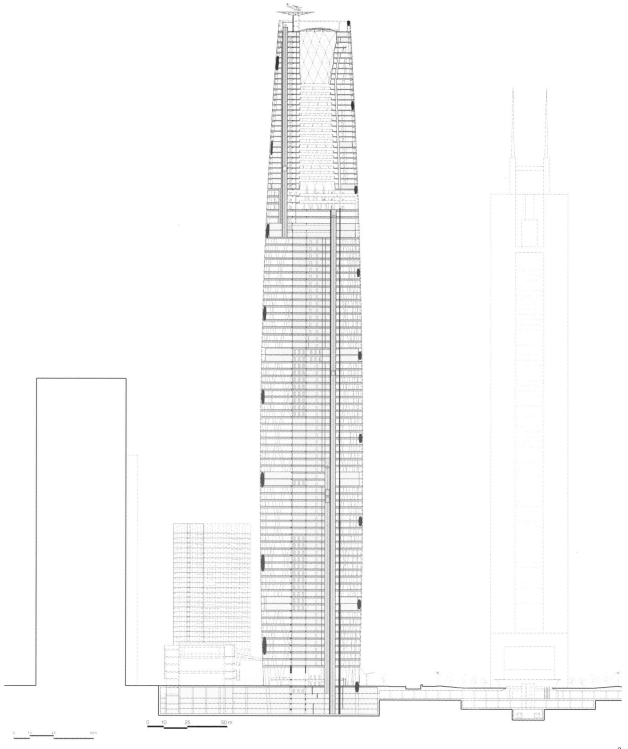

8 剖面

GUANGZHOU EAST TOWER
广州东塔

威尔金森·艾尔建筑事务所 | Wilkinson Eyre Architects

项目客户：越秀投资集团
建筑设计：威尔金森·艾尔建筑事务所
工程设计：合乐 Yolles
竞　　赛：2008年9月
总用地面积：43.9 hm²
办公建筑面积：28.5万 m²
建筑层数：105层
建筑高度：475 m

在西塔项目成功设计并开始现场施工后，广州市亦启动了东塔项目的国际土地招标。政府希望各开发商在制定新东塔设计时以尊重威尔金森·艾尔建筑事务所的原双塔设计概念为前提，并符合原来城市规划的理念。同时，广州市亦接受开发方以商业理由发展自己独特的设计方案。

威尔金森·艾尔的广州西塔设计客户——越秀集团，邀请我们为其承担东塔设计的投标。设计方案遂以尊重广州规划局的意图和原双塔设计概念为原则来精心设计，因此，东塔的方案起始于对西塔的补充，同时发展自己独特的风格。

设计一开始遵循着西塔的形式，以三边平面为基础，并加以独特的曲线外观。不过，东塔的高度设置为475 m，而西塔是440 m，由于建筑物的曲率关系到细节的设计和研究，要在保证建筑高度增加的同时，维持建筑外观的清晰轮廓。

东塔的设计减少了西塔中复杂的区域，原塔中的三个曲面"节点"让建筑物的三个面融合为一个无缝、平滑的三维几何形式，这带来设计的复杂性。为此，三个复杂的曲面"节点"被横切而去掉弧度，简化了立面不必要的复杂性。同时，为了加快施工速度，复杂的三维直径网结构被省略，取而代之的是一对巨型支柱和悬臂梁结构，并于塔楼的核心筒处相连。带状桁架在设备层与巨型支柱相连，使下层的办公空间从支柱中解放出来，使其成为办公区域顶部的理想的行政空间。

105层的塔楼以办公空间为主，公共观光层和餐厅位于高楼顶部。一系列高达两层的空中花园穿插于每15层楼之间，这里也是双层电梯的转搭层，这些空中花园赋予塔楼螺旋形，在不同角度提供壮观的城市景观。一体化的空间花园在塔楼中，不但给办公区域提供了宝贵的"过渡空间"，也是对广州作为花园城市象征的一个垂直诠释。

塔的设计中尽量减少能源消耗，并将垂直轴风力发电机和光伏电池板整合于主要的观光楼层和设备楼层中。双幕墙结构也被运用在主要的立面上。

裙房形态极具动态曲线元素，包含了多达30 000 m²的商铺面积。同时塔楼和裙楼通过一个连续曲面的玻璃中庭连接。该裙楼连接到地下商铺空间，进而连通东，西塔，提供一个无缝连接的地下商业环境。

After the West tower project was successfully designed and the project had begun construction on site, the City of Guangzhou launched an international land tender for the East tower project. The city required that participating developers should develop designs for the new East tower which respected Wilkinson Eyre's original twin tower competition winning concept, and therefore their original planning parameters. However, the city of Guangzhou accepted that for commercial reasons all participating developers would want to develop their own unique design proposals.

Wilkinson Eyre's client for the west tower, the Yuexiu group, invited them to develop designs on their behalf for the East tower. Our design proposal was carefully developed to respect the intent of the Guangzhou planning authority and our original design for the twin towers. Our proposal therefore was complimentary in design to the west tower but possessing of its own unique character.

The design shares both the 3 sided plan as the west tower and its a distinctive curving external form. However as the design height

高层建筑 | 设计作品

1 建筑全景

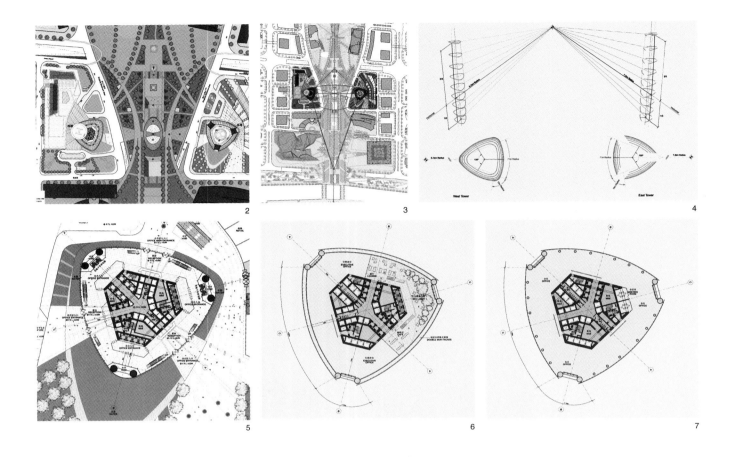

for the East tower was set at 475m and the West Tower was 440m, the curvature of the building was the subject of detailed design studies and analysis in order to allow for the required extra height whilst retaining a complimentary building profile.

The design of the tower was further developed by reducing areas of complexity apparent in the design of the West Tower. The original tower had 3 curved 'corners' which allows the 3 sides of the building to flow together into a seamless, smooth, 3 dimensional form, but this had introduced some complexity into the overall design of the building. The three 'complex' corners were therefore resolved to a chamfered detail with the rounded 'noses' removed, which in turn simplified the cladding design.

In order to simplify structural connections and speed up future construction, the complex 3D geometry of the diagrid structure was omitted. In its place a twinned mega column and outrigger system was adopted, which linked back to the buildings structural core. Belt trusses then tie the mega columns together and are located at plant floors up the vertical height of the building. This allowed office floor plates below the plant rooms to be 'column free' and ideal for use as executive floors located at the top of each office zone.

The 105 storey tower contains predominately office space with restaurant and public sightseeing levels located at the top of the building. A series of dramatic double height 'sky gardens' are located every 15 storeys, at double decker lift transfer floors and spiral up the shape of the tower, offering spectacular changing views of the city. The integration of gardens throughout the tower will provide valuable 'transient space' in the working environment and a vertical interpretation of the garden city from which it rises.

The tower as been designed to minimise energy consumption and integrates highly visible vertical axis wind turbines and photovoltaic's cells at plant floors and at the main sightseeing level. Double skin facades have also been integrated on key elevations.

A podium contains a further 30,000sqm of retail and office space contained within a dynamic curving element and expressed mini-tower. A continuous curving glazed atrium links the tower with the podium. The podium connects to below-ground retail space which in turn links both East and West Towers to provide a seamless below ground retail environment.

2 总平面
3 场地构成分析
4 体量关系分析
5 首层平面
6 办公空间与休闲空间的配置
7 办公空间与多功能会议空间

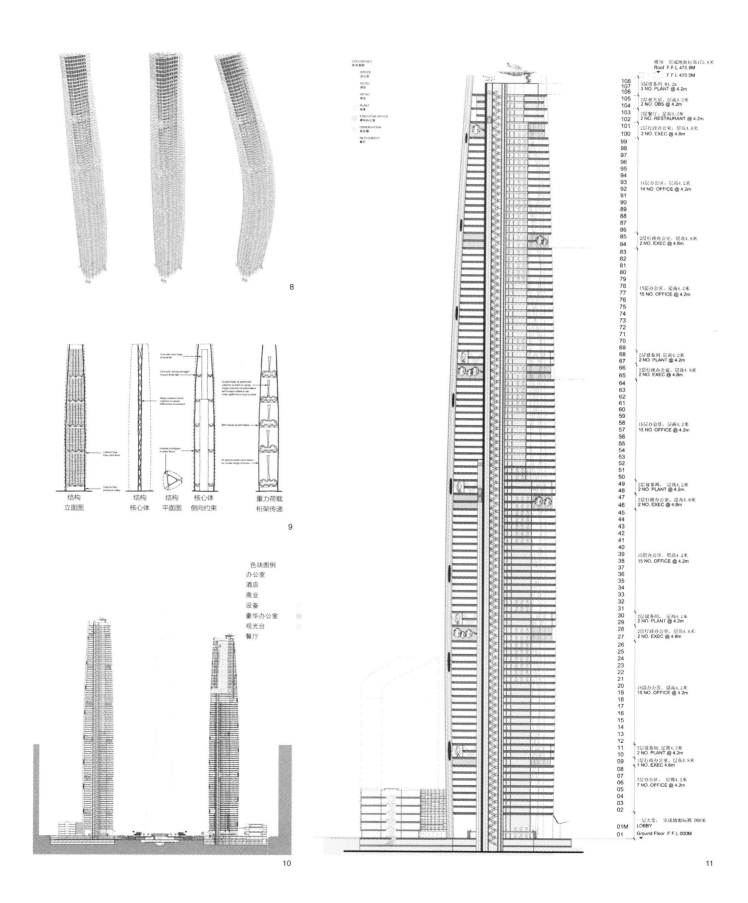

8 形体分析
9 结构分析
10 东塔与西塔剖面
11 东塔剖面

THE TADAWUL STOCK EXCHANGE TOWER, KINGDOM OF SAUDI ARABIA
沙特阿拉伯 Tadawul 证券交易大楼

威尔金森·艾尔建筑事务所 | Wilkinson Eyre Architects

项目客户：Tadawul
建筑设计：威尔金森·艾尔建筑事务所
工程设计：WSP Group
建筑面积：14.4 hm²

在利雅得城新规划的阿卜杜拉国王金融区，威尔金森·艾尔建筑事务所在沙特阿拉伯证券交易所大楼设计（Tadawul）的竞标中，提出了一个独特的建筑方案，隐喻了历史悠久的阿拉伯文化和当代的技术经济文化。这座48层、高200 m的低碳环保办公大楼作为新金融区的一部分，将给利雅得城带来14.5万 m²的优越办公环境。

在业主的要求当中，一个关键的挑战就是如何建立一个具有里程碑意义的总部大楼，在满足功能要求的同时也要考虑总体规划对大楼高度的限制。威尔金森·艾尔建筑事务所的方案是在对未来该区域建设的大量分析后，综合考虑与新金融区内总部大楼周围建筑物关系的基础上提出的。

建筑群平面的自然断裂形式，是参照传统阿拉伯马赛克的几何形式而创造的，以三角几何图案有机组成结晶式图案。该晶体的几何形式在视觉上有助于分散并减弱大型建筑物的体量，当它们被运用到外墙的立面并创造出一系列垂直细长条状图案的时候，更进一步减轻了建筑物的体积感。垂直的立面被这种螺旋的几何形式进一步解构并重组，形成精巧的菱形结构外墙。不管从哪个角度去观察，每个顶点随着三角形折面或凸起或凹陷，都会形成一个充满动感的"地形倾斜"的玻璃幕墙结构外墙立体外墙结构。交替倾斜的平面衬托出了建筑物的垂直感和轻盈感，并增添了趣味性。

大楼的特征之一是塔和平台之间的一个遮阳结构，它将"外部"包装为一个连续的形式联系着一系列平面。这种遮阳栅栏采用起伏形式将东部和西部的塔楼外墙包裹起来。这个简单的构件激起了人们对沙漠被风吹过后产生涟漪的想象，让整个大楼给人留下鲜明的印象。金色的百叶窗更进一步地加强了这种效果，随着楼层的增高，金黄色渐渐褪去成为银色，反射出天空的色彩，与蓝天相接。

利雅得的名字来源于阿拉伯语"rowdhah"，意思是"花园之地"。贯穿整座大楼的花园和水景的一体化设计体现了自然与旧城居住共存的气氛，给予了现代建筑生态技术最全面的诠释。

建筑师具体地考虑到每个表面不同的环境条件，创造出随朝向而改变的建筑形式。早期的动态三维计算机模拟优化了外观设计，减少了热能损失，同时提供最大限度的遮阳效果，从而减少眩光，保持适当的采光。尖脊造型的外观设计进一步加强了遮阳的效果。

陡峭的遮阳栅栏，在建筑表面上共同作用，起到了最大的遮阳效果，并且解决了建筑内外的沙尘问题。遮阳栅栏设计中的微妙变化无论在远距离还是近距离，都能创造出波光粼粼的立面视觉效果和栩栩如生的外观。

北立面和南立面的外观类似于水晶被切割打磨后的形式，而东立面和西立面形成了纹理粗糙的边缘。为了最大限度地获得贯穿利雅得城的视野，南立面保持了非常高的透明度，同时利用双幕墙技术缓解了当地酷热的气候条件。

位于大楼下方的两层楼提供了超过6 600 m²的高科技交易厅，由4个数据中心组成，为沙特证券交易所提供了一个电子交易平台。而媒体中心的电视广播工作室服务于利雅得的金融电视台，包括一个电视直播室和一个500座席的会议中心。该设计确保了利雅得的总部大楼将成为世界上公认的最高效和具备先进技术的证券交易所之一。

在证券交易所上方可租用的办公空间，被设计为高效、灵活、易于分隔以及不同尺寸的办公室，这样更能满足灵活的租赁市场。

风力涡轮发电机安置于每个角落绿地中，为整个大楼提供"绿色能源"。此外安置于楼顶的太阳能光伏阵列也可提供电力能源。

1 建筑入口

Located within Riyadh's newly planned King Abdullah Financial District, Wilkinson Eyre's competition design for The Saudi Arabian Stock Exchange tower, or Tadawul, proposes a unique architectural statement that is suggestive of both Arabic and digital economic culture. This 48 storey, 200m high, low energy office tower forms part of an ambitious new financial district and will provide 145,000 sqm of premium office floor space for the Tadawul.

One of the key challenges of the brief was to create a landmark headquarters building that met a very ambitious floor space requirement whilst observing the height restrictions imposed by the masterplan. The massing of the scheme is based on an analysis of the proposed context, and more specifically the massing of the surrounding buildings that will form the new financial plaza.

The plan form was fractured, creating a crystalline pattern of triangular forms with references to traditional Arabic mosaic geometry. The crystalline geometry helps to fragment and reduce the visual mass of the building; it is also transposed onto the vertical planes of the facades, creating a series of slender vertical strips which further break up the perceived mass. The vertical planes of the facade are further articulated by a spiral geometry, resulting in triangulated diamond shaped facets to the facades. The apex of each triangular fold is either raised or recessed in order to create a dynamic "topography" of inclined and declined glazing, when looking up or across any facade of the building. The alternating inclination of the surface planes means that various surfaces of the building will catch the light differently, reinforcing the sense of verticality and lightness and providing interest.

One of the defining characteristics of the development is the visual and physical link between the tower and podium in the form of a continuous plane of sunshade louvres that wraps over the 'external' faces of the building as a series of ribbons. This undulating plane of louvres forms the east and west facades of the tower and wraps up and over the high level and podium roofs in a continuous sweep creating shaded roof terraces. This simple device provides a powerful identity for the whole development reminiscent of the windblown ripples of the desert. This analogy is further enhanced by the golden finish of the louvers at the base relating to colour of the desert which fade to silver towards the top of the tower reflecting the colour of the sky. Riyadh's name is derived from the Arabic word 'rowdhah,' meaning 'place of gardens'. The tower integrates numerous gardens, planted terraces and water features in a design which seeks to reflect the former atmosphere of the Old City. This vertical reinterpretation of the 'rowdhah,' provides a highly visible statement of ecological technology and the best modern working practices.

The form of the building changes with orientation and responds specifically to the very different environmental conditions that each facade experiences. Early stage dynamic 3D computer simulations optimised the facades; to minimise heat loss, maximise solar shading and reduce glare, whilst maintaining appropriate daylighting. The topography of the facade has been designed to be self shading with the ridges provide further shading to the facade.

The steep angle, profile and finish of the louvres maximise solar shading whilst preventing sand and dust from settling on its surface. It is intended that the variation and subtlety in the cladding design will create shimmering elevations enlivening the facades from both long distance and close up views.

The North and South facades have come to resemble the cut, smooth surfaces of the crystalline form with the East and West facades forming the textured rough edges. In order to maximize views across Riyadh, the South facade maintains a high level of transparency and solar performance in this demanding climate by utilizing a double skin facade technology.

Within the building, the lower two floors provide over 6,600sqm of high tech trading floors. The trading floors are served by a state of the art Tier 4 Data Centre which provides the electronic trading platform for the Saudi stock exchange. A media centre provides TV studios for Tadawul's financial TV station providing live broadcasting facilities with associated conference centre and 500 seat auditorium. The design ensures that Tadawul's Headquarters will be internationally recognized as one of the most efficient and technologically advanced stock exchanges in the world.

The lettable office space above the stock exchange has been designed to be highly efficient and flexible, with easily sub-dividable floor plates. The leased offices provide a range of tenancy sizes to give flexibility of leasing and market requirements.

No. wind turbines located at each corner of the plant levels provide 'green energy' for the tower. Electricity is also be derived from photovoltaic solar arrays, incorporated within the louvre panels and glazing at roof levels.

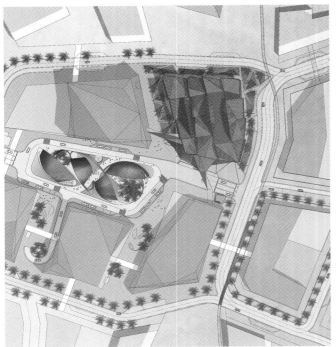

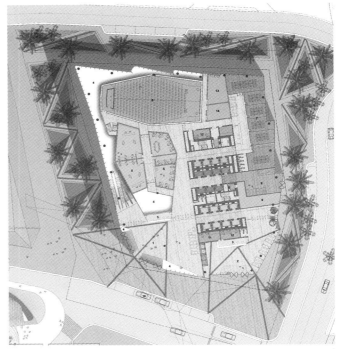

3

4

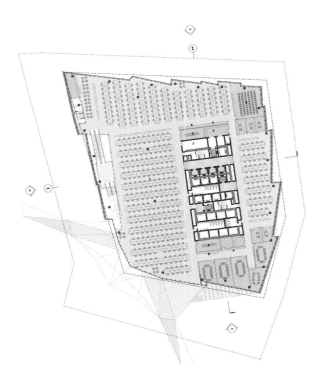

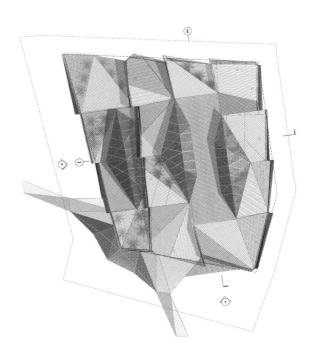

3 总平面
4 首层平面
5 二层平面
6 顶平面

5

6

7 剖面

55 DEGREES ROTATING TOWER, UAE
阿拉伯联合酋长国 55°旋转塔

威尔金森·艾尔建筑事务所 | Wilkinson Eyre Architects

55°旋转塔高335 m，位于迪拜扩张规划中的朱美拉花园城市中心地带。其在花园城市中重要的地理位置与独特的设计理念将确保该项目成为新城市乃至整个迪拜的焦点。

威尔金森·艾尔建筑事务所根据客户的要求，构想出一个每周旋转一圈的塔楼建造计划，将同世界其他大都市联合起来建造一系列塔楼。这个位于纬度55°的迪拜塔楼，将是该计划的第一座，之后亦会在伦敦、香港和纽约陆续建造。

塔楼被定位为以居住用途为主的混合使用项目，高档办公空间被安排在塔楼低层，高端餐厅及酒吧则被设置在塔楼的顶端。塔楼被设想为有生命的建筑，慢慢地、连续不断地在一个星期的时间内完成一周的"进化"。

我们将建筑极具雕塑感的多边形几何形式与旋转的天数巧妙地结合起来，使其每天缓慢的运动也能够被整个城市的人们察觉到。这些公寓将绕着固定的核心筒旋转，远方的景色映入眼帘，居住在旋转塔里的人能够在每天一早起来都面对一个新的景色。

在结构上，成功地让一个庞大的结构（重量几十万吨）旋转的技术难题是由我们的合作工程师奥雅纳解决的。奥雅纳采用了最先由美国宇航局为航天飞机设计的轴承技术，使该建筑围绕着升降机和逃生楼梯构成的核心筒旋转。固定服务的出入问题也通过一些创新技术的应用得到解决。

横向的遮阳结构设置于倾斜向上的建筑立面，给予居住者庇护，以应对迪拜酷热的环境。而倾斜向下的立面则采用垂直的遮阳结构。遮阳栅栏采用金色、银色和铜色不同涂漆，随着旋转的建筑立面而变得更加生动。

夜晚，塔楼因灯火而变得晶莹剔透，建筑结构的边线和节点也被生动地表现出来。

The 335m, '55 degrees Rotating Tower' forms the centre piece for Jumeirah Garden City, an ambitious new masterplan for the expansion of Dubai. The towers prominent location within the Garden City and unique design will ensure that it will become a destination for this new urban quarter and for Dubai as a whole.

Wilkinson Eyre were presented by our client, with the idea of creating a tower which completed a single revolution every week and which would form a series of towers located in every major city across the world. Dubai, located on 55 degrees latitude, would be the first tower to be designed and constructed and would be followed by London, Hong Kong and New York.

The brief required a mixed use tower of predominantly residential use but with boutique offices at lower levels and a high end restaurant, bar and a club located at the apex of the building. The tower was required to slowly and continuously rotate, completing one revolution per week and is conceived as a "living" building which is constantly transforming.

The highly sculptural shape, which utilises an underlying 7 sided heptagonal form relating poetically to the number of days in the week, allows the slow daily movement of the tower to be fully appreciated across the city as the shape seamlessly shifts and evolves. The residential apartments rotate around a fixed core, which are treated to far-reaching views which constantly replenish, awakening each morning to a fresh vista.

The huge technical challenge of allowing a structure weighing several hundred thousand tonnes to rotate was addressed by engi-

1 临街效果
2 日景效果
3 夜景效果

高层建筑 | 设计作品

151

4

neers Arup who adopted bearing technology first developed for the NASA space shuttle. The building rotates around a fixed core containing lifts and escape stairs. The challenge of overcoming fixed services flowing into and out of a rotating building was also over come using several novel techniques.

Horizontal louvres are used to shelter apartments from the harsh Dubai sun on vertical and upward facing facets of the facade. Where the facet of the facade leans downwards and is sheltered by the building above, vertical louvres are used. By using different polished gold, silver and bronze finishes on each side of the louvres, the facade is further animated as it rotates.

By night the building turns 'inside out': the seams of the structure and the node points are illuminated.

4 充满动感的体量构成

高层建筑 | 设计作品

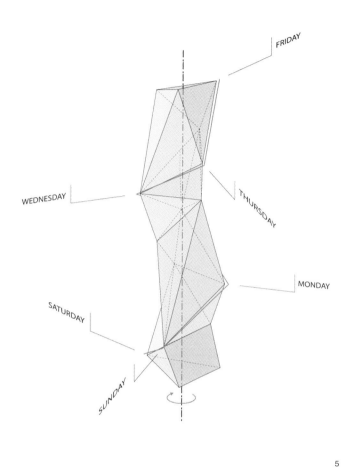

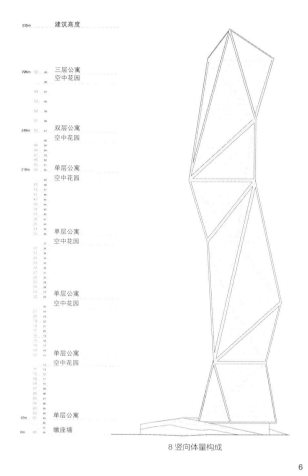

8 竖向体量构成

5　　　　　　　　　　　　　　　　　　　　　　　　　　　　　　　　　　　　　　6

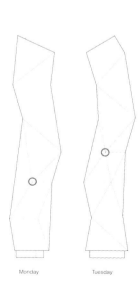

Monday　　Tuesday　　Wednesday　　Thursday　　Friday　　Saturday　　Sunday

The faceted shape means that the facade is constantly shifting and changing as it rotates

5　体量的分时段旋转
6　竖向体量构成
7　旋转形态

7

SHENZHEN STOCK EXCHANGE HEADQUARTERS
深圳证券交易所新总部大楼

大都会建筑事务所 | Office for Metropolitan Architecture (OMA)

项目名称：深圳证券交易所新总部大楼
业　　主：深圳证券交易所
建设地点：深圳市滨河大道
用地面积：3.9 hm²
建筑面积：26.5万 m²（地上18万 m²，地下8.5万 m²）
建筑层数：46层
建筑高度：254 m
设计单位：OMA
主管合伙人：Rem Koolhaas, David Gianotten
协同合伙人：Ellen van Loon, Shohei Shigematsu
协理建筑师：Michael Kokora
现场团队：Yang Yang, Wanyu He, Daan Ooievaar, Joanna Gu, Vincent Kersten, Yun Zhang
设计团队：Kunle Adeyemi, Ryann Aoukar, Sebastian Appl, Laura Baird, Waichuen Chan, Jan Dechow, Lukas Drasnar, Matthew Engele, Leo Ferretto, Clarisa Garcia Fresco, Alasdair Graham, Jaitian Gu, Matthew Haseltine, João Ferreira Marques Jesus, Matthew Jull, Alex de Jong, Santiago Hierro Kennedy, Klaas Kresse, Miranda Lee, Anna Little, Luxiang Liu, David Eugin Moon, Cristina Murphy, Se Yoon Park, Ferdjan Van der Pijl, Franscesca Portesine, Idrees Rasouli, Korbinian Schneider, Wolfgang Schwarzwalder, Felix Schwimmer, Richard Sharam, Lukasz Skalec, Christine Svensson, Lukasz Szlachcic, Ken Yang Tan, Michela Tonus, Miroslav Vavrina, Na Wei, Xinyuan Wang, Leonie Wenz, Su Xia, Yunchao Xu, Yang Yang, Yun Zhang
竞赛团队：Konstantin August, Andrea Bertassi, Joao Bravo da Costa, Tieying Fang, Pei Feng, Katharina Gerlach, Carlos Garcia Gonzalez, Martti Kalliala, Klaas Kresse, Anu Leinonen, Anna Little, Jason Long, Beatriz MInguez de Molina, Daniel Ostrowski, Yuanzhen Ou, Mauro Paraviccini, Mendel Robbers, Mariano Sagasta, Bart Schoonderbeek, Hiromasa Shirai, Kengo Skorick, Hong Yong Sook, Christin Svensson, Xinyuan Wang, Dongmei Yao
声学设计：Bertie van de Braak, Caroline Kaas, Renz van Luxemburg, Theo Rijmakers (DHV Building and Industry)
景观设计：Petra Blaisse, Rosetta Elkin, Aura Melis, Jana Crepon with Laura Baird and Carmen Buitenhuis
结构/设备/消防/项目管理/垂直交通/楼宇智能/地质技术/照明设计：Arup
当地设计院设计人员：Yuan Chao, Jing Chen, Jun Chen, Wen Deng, Bo Hong, James Hong, Zhen Hu, Ming Huang, Hanguo Li, Wenming Lin, Zhenhai Lin, Chen Liu, Qiongxiang Liu, Jianlin Mao, Jianmin Meng, Zhijian Qiu, Xiaoheng Shen, Xingliang Shi, Luming Shu, Nan Sun, Xiaohong Sun, Qiwen Wang, Yishan Wang, Chao Wu, Fenghua Xiao, Chuangui Xie, Baozhen Yang, He Yang, Hui Zhen, Wenxing Zhen（深圳建筑设计研究院）
建设时间：2006年
建成时间：2013年10月
图纸版权：OMA
摄　　影：Philippe Ruault

　　证券市场的本质是投机：它基于资本，而非物质。深圳证券交易所新总部大楼是虚拟证券市场的物化，这座有着悬浮基座的建筑象征着证券市场的本质，而不是仅容纳其功能。通常情况下，建筑的基座支撑着建筑的结构并将其与地面紧紧相连。然而，深圳证券交易所新总部大楼却与众不同，它的基座仿佛被推动市场的乐观的投机情绪所抬高，上升到塔楼，成为升起的裙房，挑战存活到现代的千年建筑传统——坚实建筑立于坚实的基座之上。

　　建筑将3层的基座抬升至距离地面36 m的高度，构成悬挑平台，形成面积最大的办公楼层（每层面积为1.5万 m²），同时建构出可上人的屋顶花园。抬升的基座容纳了证交所所有的功能，包括上市大厅以及所有办公部门。建筑的基座因其抬升的高度大大提升了深交所的辨识度。建筑在夜晚灯火通明，仿佛向整个城市"广播"着深圳金融市场的活动，悬挑的平台勾勒出深圳的景色。此外，基座的抬升也释放了地面的空间，为安全、私密的建筑创建了宽敞的公共空间。

　　抬升的基座与塔楼为整体结构，塔楼和中庭的支柱为悬挑结构提供竖向和横向的支撑。抬升的基座由坚固的三维全深度钢转换桁架为框架。塔楼的两侧设计了两个将地面与建筑内部公共空间直接相连的中庭。深交所的工作人员从东侧进入大楼，而租户从西侧进入。深交

1 深交所正方形的塔楼外观使其与周边的高楼很好地融合在一起，但因其抬升的基座大大提高了建筑的辨识度

所的行政办公室位于基座的上层，再上面的楼层是可租赁的写字楼和餐饮会所。

深交所正方形的塔楼外观使其与周边的高楼很好地融合在一起，但其外立面却独树一帜。压花玻璃包覆着支撑建筑的网格结构骨架，表皮的肌理反映出建筑背后的建造技术，同时使建筑散发出一种神秘的美感。立面中性的色彩及透明度随着天气而变化，创造出神秘的水晶般的效果：晴天闪耀，阴天暗淡，黄昏闪亮，夜晚璀璨。建筑"深邃的立面"上凹进的开口减少了太阳辐射热量的进入，同时又提高了自然采光效果，减少了能耗。深圳证券交易所新总部大楼将成为中国首批达到绿色建筑三星标准的大楼。

这座46层（254 m高）的建筑是一栋富于城市意义的金融中心，它位于深圳市莲花山和滨河大道的南北轴线与深圳主要干道深南大道东西轴线交汇处的新的公共广场。它并不是一栋孤立的建筑，而是在多个层面上呼应了城市，既表现出恢宏，又不失亲切，不断地与城市建立新的关系，希望借此推动有关建筑和城市的新形式的思考。

The essence of the stock market is speculation: it is based on capital, not material. The Shenzhen Stock Exchange is conceived as a physical materialization of the virtual stock market: it is a building with a floating base, representing the stock market – more than physically accommodating it. Typically, the base of a building anchors a structure and connects it emphatically to the ground. In the case of Shenzhen Stock Exchange, the base, as if lifted by the same speculative euphoria that drives the market, has crept up the tower to become a raised podium, defying an architectural convention that has survived millennia into modernity: a solid building standing on a solid base.

SZSE's raised podium is a three-storey cantilevered platform floating 36m above the ground, one of the largest office floor plates, with an area of 15,000 m² per floor and an accessible landscaped roof. The raised podium contains all the Stock Exchange functions, including the listing hall and all stock exchange departments. The raised

2 建筑在夜晚灯火通明，仿佛向整个城市"广播"着深圳金融市场的活动，悬挑的平台勾勒出深圳的景色

podium vastly increases SZSE's exposure in its elevated position. When glowing at night, it "broadcasts" the virtual activities of the city's financial market, while its cantilevers crop and frame views of Shenzhen. The raised podium also liberates the ground level and creates a generous public space for what could have been what is typically a secure, private building.

The raised podium and the tower are combined as one structure, with the tower and atrium columns providing vertical and lateral support for the cantilevering structure. The raised podium is framed by a robust three-dimensional array of full-depth steel transfer trusses. The tower is flanked by two atria – voids that connect the ground directly with the public spaces inside the building. SZSE staff enter from the East and tenants from the West. SZSE executive offices are located just above the raised podium, leaving the uppermost floors leasable as rental offices and a dining club.

The generic square form of the tower obediently blends in with the surrounding homogenous towers, but the facade of SZSE is different. The building's facade wraps the robust exoskeletal grid structure supporting the building in patterned glass. The texture of the glass cladding reveals the construction technology behind while simultaneously rendering it mysterious and beautiful. The neutral colour and translucency of the facade change with weather conditions, creating a mysterious crystalline effect: sparkling during bright sunshine, mute on an overcast day, radiant at dusk, and glowing at night. The facade is a "deep facade", with recessed openings that passively reduce the amount of solar heat gain entering the building, improve natural day light, and reduce energy consumption. SZSE is designed to be one of the first 3-star green rated buildings in China.

The 46-storey (254m) Shenzhen Stock Exchange is a Financial Center with civic meaning. Located in a new public square at the meeting point of the north-south axis between Mount Lianhua and Binhe Boulevard, and the east-west axis of Shennan Road, Shenzhen's main artery, it engages the city not as an isolated object, but as a building to be reacted to at multiple scales and levels. At times appearing massive and at others intimate and personal, SZSE constantly generates new relationships within the urban context, hopefully as an impetus to new forms of architecture and urbanism.

3　抬升的基座构建出可上人的屋顶花园

HIGH-RISE BUILDING | DESIGN WORKS

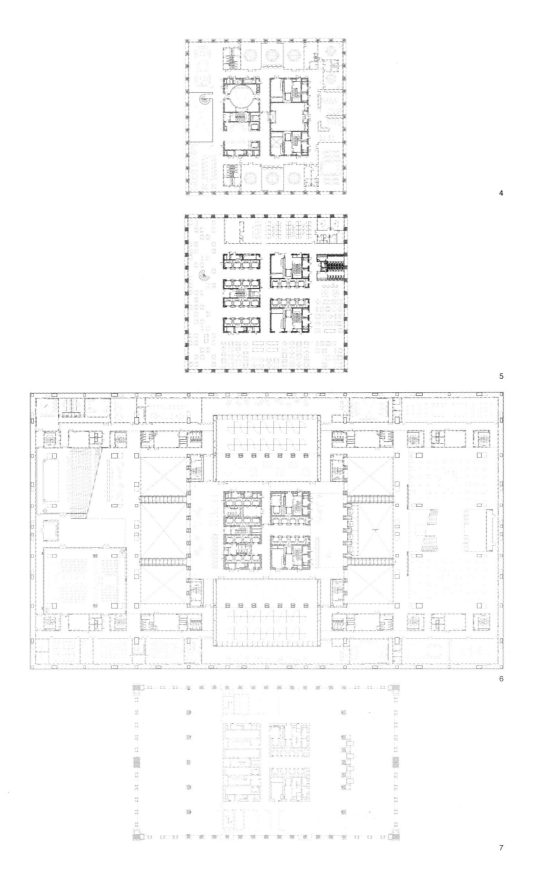

4 四十六层平面
5 三十三层平面
6 八层平面
7 一层平面

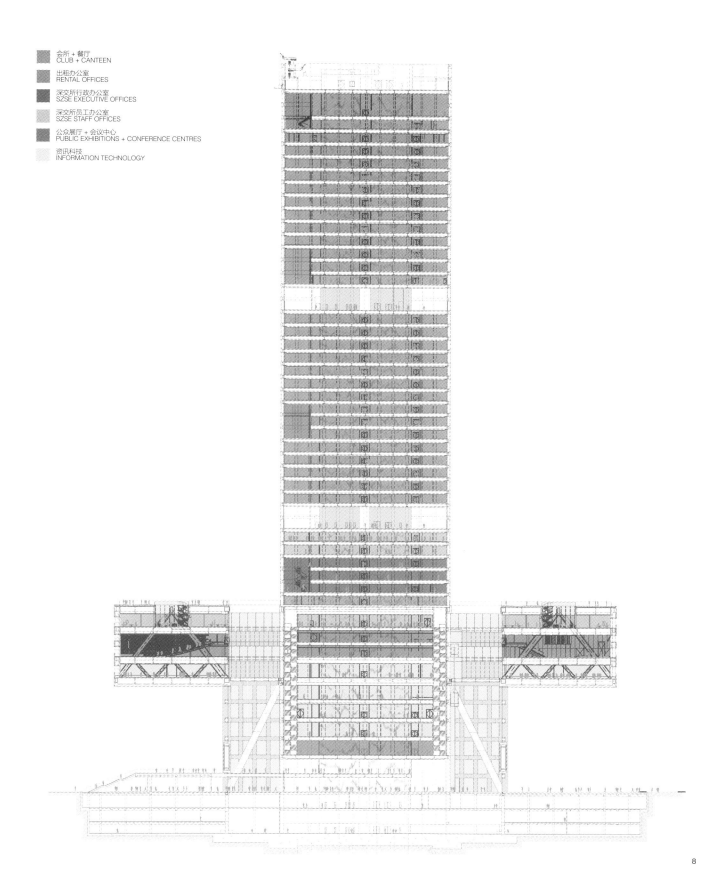

	会所 + 餐厅 CLUB + CANTEEN
	出租办公室 RENTAL OFFICES
	深交所行政办公室 SZSE EXECUTIVE OFFICES
	深交所员工办公室 SZSE STAFF OFFICES
	公众展厅 + 会议中心 PUBLIC EXHIBITIONS + CONFERENCE CENTRES
	资讯科技 INFORMATION TECHNOLOGY

8 剖面

9

10

9 建筑将3层的基座抬升至距离地面36 m的高度,构成悬挑平台
10 基座的抬升释放了地面的空间,为安全、私密的建筑创建了宽敞的公共空间

11

11 层次丰富的内庭空间
12 凹进的开口在夜色中的效果
13 建筑"深邃的立面"上凹进的开口减少了太阳辐射热量的进入，同时又提高了自然采光效果

12　　　　　　　　　　　13

111 FIRST STREET, USA
美国 111 第一大街

大都会建筑事务所 | Office for Metropolitan Architecture（OMA）

项目业主：BLDG Management Co. Inc. and The Athena Group, LLC
项目规模：总建筑面积——1.2 million SF，公寓面积——415 000 SF，酒店面积——210 000 SF，艺术工作室/直播工作室面积——160 000 SF，画廊面积——19 000 SF，商业零售面积——87 000 SF，停车面积——240 000 SF
建筑高度：592 ft
项目主创：Rem Koolhaas
协助负责：Shohei Shigematsu
设计团队：Noah Shepherd, Christin Svensson, Margaret Arbanas, Kengo Skorick, Alasdair Graham with Torsten Schroeder, Chun Yue Chiu, Duncan Flemington, Martin Schliefer, Tomek Bartczak, Javier Munoz, Ian Robertson and Richard Hollington, Simon de Jong, Jan Dechow, Adam Frampton, Ludwig Godefroy, Jin Hong Jeon, Klaas Kresse, Jang Hwan Lee, Jung Hwan Park, Jesse Seegers
建筑撰文：SLCE Architects-Peter Claman, Robert J. Laudenschlager
建筑结构：WSP Cantor Seinuk Structural Engineers-Silvian Marcus, Patrick Chan, Jay Harris, Susan Hamos
MEP：Flack+Kurtz Consulting Engineers-Daniel Nall, Andrew Hushko, Robert Sedlak
建筑效果：Frans Parthesius / OMA
模型制作：OMA/Made by Mistake

发展中心

111第一大街隶属美国泽西市新兴海滨开发区，坐落于由火车站、轮渡站、隧道及直升机场组成的公共交通枢纽中心区。总建筑面积11.15万 m²（公寓3.85万 m²，宾馆及辅助设施1.95万 m²，艺术家住宅兼工作室1.49万 m²，画廊0.17万 m²，零售店0.8万 m²，停车场2.22万 m²），这种功能组合模式将与原有建筑一起影响整个电厂艺术区，使其成为泽西市文化中心。

探索已知

我们探求如何在极限垂直状态下展现多功能组合的开发模式，并充分利用这种状态，将其转化为有利条件；如何利用类型学创造城市活化剂，而不再重复平庸；如何利用已知创造未知……

垂直城市

在111第一大街方案中，每项功能均进行了优化布局研究，并细化到每个建筑块，从艺术家住宅、工作室和画廊的每个隔间，到宾馆房间和公寓的每层楼板，再到底层公寓单元更宽的楼板。依次类推，各建筑块垂直叠加，构筑52层高（180 m）的塔楼。这种层叠结构既能使每个建筑块保持相对独立，又能获得最佳视野，使大楼与周围建筑形成动态的对话关系——在常规中构筑奇景。

建筑功能块之下的方形底座连接基地，其中商店、大堂和停车场将大楼与周围环境融为一体。随着建筑功能块方位的变换，其连接处就形成了开放空间，包括5层的111第一大街公共阳台（5 200 m²）、17层的酒店餐厅和温泉疗养所平台（1 450 m²），还有36层两块公用居民阳台（1 200 m²）。每个平台附近又设有公共活动场所，有的在白天开放（画廊、温泉疗养、体育馆、游泳池、餐厅），有的在晚上开放（歌舞厅、酒吧、餐厅、休闲室）。总之，一个中心结构贯穿三个建筑功能块，使建筑整体更加稳定，并在此设置公用楼梯直通上层。

111第一大街的公用阳台直接和街道相连，配合规划中的电厂娱乐中心和北面雕刻公园，丰富了街道的生活空间。在垂直和水平方向上所形成的密集公众活动区将带动周边地区的发展，为泽西市创造新的文化中心。

Center of Gravity

111 First Street is located in Jersey City's burgeoning waterfront development, at the center of a public transportation network created by train stations, ferry stop, tunnel and heliport. The 1.2 million square foot development's mix of program, 415,000 SF of apartments, 210,000 SF of hotel and amenities, 160,000 SF of artist work / live studios, 19,000 SF of gallery, 87,000 SF of retail and 240,000 SF of parking will, together with the existing Powerhouse, act as a beacon for the future development of the Powerhouse Arts District into Jersey City's cultural center.

Exploration of the Known

How can this mix of programs—typically hidden within the confines of relentless verticality—be revealed, harnessed, capitalized? How can we use typologies that all too often result in repetitive banality to create an urban catalyst? How can we make unknown from known?

Vertical City

Each component of the program is analyzed for optimum layout and concentrated into individual blocks—a cube of artist work, live studios and galleries, a slab that combines hotel rooms and apartments, and a wider slab that accommodates deeper apartment units. The resulting volumes are stacked perpendicularly in plan to create a 52 story, 592 ft, tower. The stacking maintains the independence of each block, optimizes views from the site and creates a dynamic relationship between the building and its surroundings: Spectacle from Convention.

Beneath the blocks, a plinth fills the site, connecting the building to the city with a mix of retail, lobbies, and parking. Alternating the orientation the blocks create a series of open spaces at their junctions: the 111 First Street Public Terrace on the 5th Floor 56,000 SF, terraces for the hotel restaurant and spa on the17th Floor 15,700 SF and two shared residential terraces on the 36th Floor 13,000 SF. Adjacent to each terrace is a public space that activates it during the day gallery, spa/gym/pool, restaurant. and night cabaret, bar, restaurant, residential lounge. A central core joins all three blocks, giving structural stability to the building and providing access to upper public floors.

With direct street access, the 111 First Street Public Terrace will activate the street life and create a synergy between the planned Powerhouse Entertainment Center and the Sculpture Garden north of the site. The vertical and horizontal density of public activity generated will energize the surrounding area, creating a cultural hub for Jersey City.

1 建筑效果图
2 建筑体块构成分析

1　　2

HIGH-RISE BUILDING | DESIGN WORKS

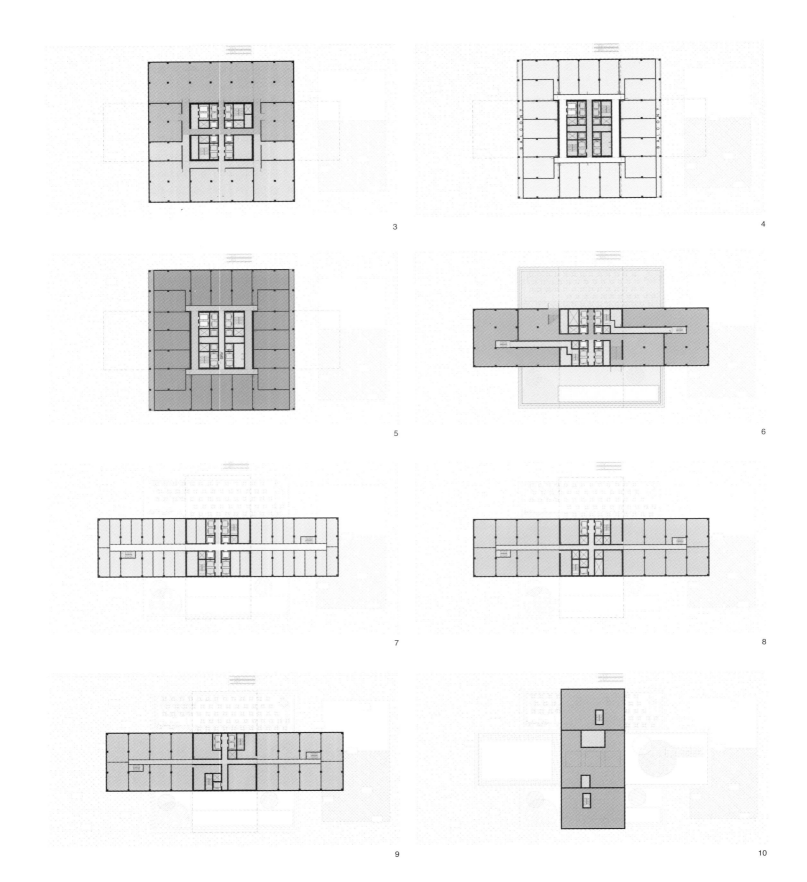

3

4

5

6

7

8

9

10

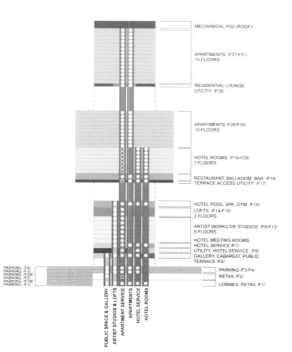

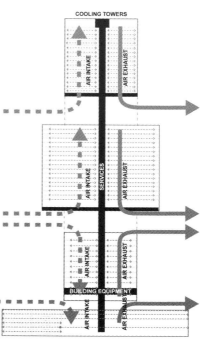

3　五层平面图
4　八至十三层平面图
5　十四至十五层平面图
6　十七层平面图
7　十九至二十五层平面图
8　二十六至二十七层平面图
9　二十八至三十五层平面图
10　五十二层平面图
11　建筑在街区中的效果
12　建筑垂直循环系统分析

LA TOUR PHARE, FRANCE
法国灯塔高楼

大都会建筑事务所 | Office for Metropolitan Architecture（OMA）

项目业主：Unibail
项目规模：165 m²，高度 300 m
项目功能：办公、屋顶停机坪、体育馆、餐饮、商业零售、停车场
项目建筑主创：Rem Koolhaas and Ellen Van Loon
项目建筑师：Clément Blanchet
设计团队：Adrianne Fisher，Alain Peauroi，Alejandro Aguirre de Carcer，Anne-Sophie Bernard，Bin Kim，Charles Antoine Perreault，Daphna Glaubert，David Nam，Duncan Flemington，Jan Dechow，Javier Munoz，Karen Crequer，Katrin Bernickel，Keigo Kobayashi，Luca Astorri，Mauro Parravicini，Michal Gdak，Oliver Luetjens，Phillipe Braun，Tomek Bartczak，Yang Yang

项目顾问：ARUP-Chris Carroll-Arup Director，Nadia Molenstra-S，William Whitby-S，Philippe Schmit-S，Julian Sutherland-MEP，Michael Bingham-MEP，Tony Lowell-Fire，Andrew Hall-Façade，Stuart Clarke-Facade
竖向交通：Lerch and Bates
停车场：WHOR Autoparksysteme
建筑表现：Auralab
建筑模型：Vincent de Rijk
造价估算：Davis Langdon
图片制作：Frans Parthesius
图解设计：Irma Boom

  La Défense是一种象征，它反映了法国毫无保留地接受现代化以及随之产生的困惑。灯塔高楼设计竞赛就充分体现了这一精髓。顾名思义，塔楼具有标志性，但却要建设在一个几乎不存在的基地底座上。另一个设计上的挑战就是要将场地划分为两个部分，以便使人行道（Passerelle de l'Arche）从此通过。

  正是基于竞赛的两点要求，我们确定的设计策略包含以下三种不同的实施方案：第一，创造一个底部；第二，重新规划其合理性；第三，向外对城市投射灯光。

  塔楼底部场地由一条高速公路和铁路线围成不规则的形状。为尽量减小塔楼底部体量，设计师把楼的第一层抬起，同时改变其正立面形态，将人行道从墙面直插进来，使格局更加紧凑。为避免过多占据La Defense的地面空间，塔楼底部只能保留必要的结构功能，其中暴露出来的楼板则呈现出透视效果。

  建筑主要的矩形部分作为高效、灵活的写字楼，包括从传统的办公格局到适应现代工作需求的开放空间，一应俱全。

  遗憾的是，建筑中最有价值的部分被隐藏了。设计师计划在空中大楼这层，将大楼设计成有机的生命体：将四个附属部分盘旋在公众场所最密集的空间顶部，使其成为独立自主的部分。这种设计促使塔楼中心形成开阔的会场空间，从这里能俯视巴黎城市全貌。当然，从巴黎仰视塔楼的侧面，它也不再是严肃而充满理性的建筑物了。

  设计师正是采用这三种策略回应竞赛要求的每个细节。由于场地条件、建造环境和地理位置等因素的限制，设计并没有采用夸张、大体量、标志性和变形等手法，当然，这些造型手法在这个项目中也很难实现。

  La Défense is emblematic of the unreserved openness of the French towards modernity, as well as their current perplexity towards its result. The La Tour Phare, or the lighthouse tower, competition is representative of this: as the name suggests its function is to act as an icon, yet it will be launched from a base which is almost nonexistent. Another challenge is the division of the site into two sections allowing the passage of the pedestrian walkway Passerelle de l'Arche.

  Each of these conditions influenced our design, and results in three different operations: 1. Creation of a base, 2. Reinstallation of rationality, 3. Projecting light outwards towards the city.

At the base, the tower emerges from an irregular site defined by a neighboring motorway and a rail link. To minimize the footprint we suspend the facade and the first floors where the passageway pierces through the hall of the building to create an intense urban experience. Instead of charging the ground of La Defense with yet another intention, we reduced the footprint to the absolute essential from which the slab emerges offering perspective sights.

The main rectangular volume of the building is devoted to office floors offering maximum efficiency and flexibility: from traditional office configurations to open spaces for contemporary work conditions.

One of the weaknesses of the skyscraper is the fact that it hides its most precious organs. At the level of the sky-lobby, we propose to radicalize the status of the skyscraper into a body with organs: four satellites hover around the main volume containing the most public functions of the building making them autonomous. In the heart of the tower this gesture liberates space for a forum with surprising views over Paris. Seen from the French Capital, the profile of the tower will never be seen as just another rational and sober tower.

These three measures allow us to respond to each of the specific questions in this competition without resigning to the irresistible logic of the exceptional, the master piece, the land mark, the contemporary transformations that could not incarnate due to the site, the work, the place.

1 建筑效果

HIGH-RISE BUILDING | DESIGN WORKS

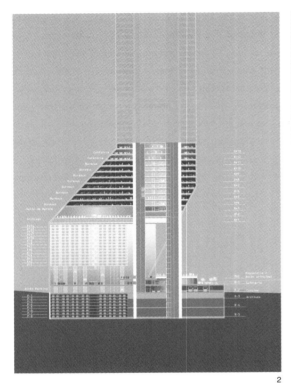

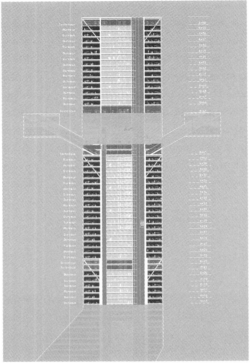

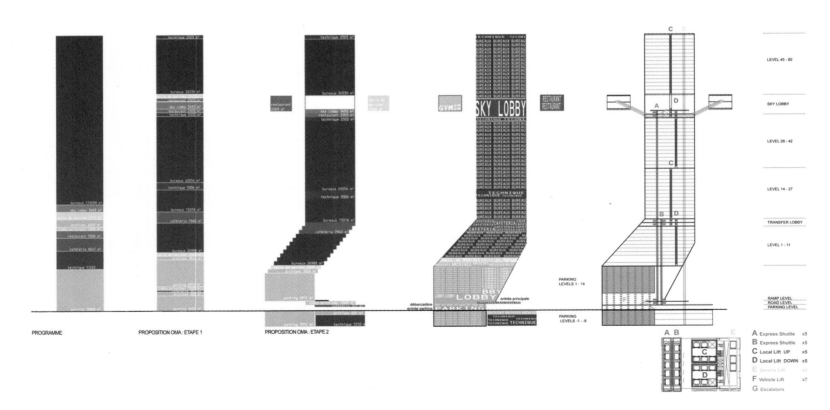

6

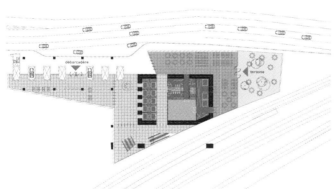

7

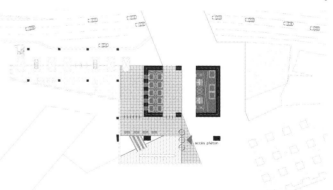

8

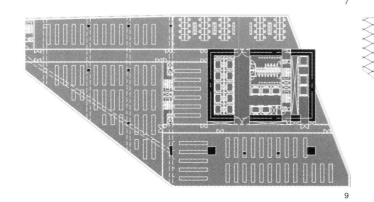

9

2　建筑低层剖面分析
3　建筑顶层剖面分析
4　建筑剖面
5　竖向垂直剖面分析
6　中庭空间模型效果
7　地下一层平面
8　一层平面
9　三层平面
10　十四层平面

TORRE BICENTENARIO, MEXICO
墨西哥 Bicentenario 塔

大都会建筑事务所 | Office for Metropolitan Architecture（OMA）

项目业主：Grupo DANHOS
项目规模：3 938 850 m²
OMA 设计团队：Rem Koolhaas，Shohei Shigematsu，Christin Svensson，Gabriela Bojalil，Noah Shepherd，Natalia Busch，Leonie Wenz，Jan Kroman，Leo Ferretto，Max Wittkopp，Jason Long，Margaret Arbanas，Jonah Gamblin，Amparo Casani，Jin Hong Jeon，Jane Mulvey，Michela Tonus，Matthew Seidel，Nobuki Ogasahara，Justin Huxol，David Jaubert，Mark Balzar，Charles Berman，James Davies，Jesse Seegers
合作建筑师：Laboratory of Architecture-Max Betancourt，Fernando Romero，Dolores Robles-Martinez
工程顾问：Arup-David Scott，Chris Carroll，Ricardo Pittella，Michael Willford，Bruce McKinlay，Julian Sutherland，Alistair Guthrie，Huseyin Darama，Yuvaraj Saravanan，Betsy Price，Keith Frankllin，Matt Clarke，Renee Mackay-Lyons
垂直交通顾问：Van Deusen & Associates-Ahmet Tanyeri
模　　型：Vincent de Rijk
说　　明：Irma Boom Design-Irma Boom，Sonja Haller
摄　　影：Frans Parthesius

　　1810年墨西哥脱离西班牙独立，100多年后，墨西哥大革命开启其现代政治的序幕。当然，一个世纪后的墨西哥、墨西哥城必将再次面对时代所带来的飞跃——全球化背景下经济和文化的繁荣必将促动墨西哥发生改变。

　　相对于世界其他区域经济发展的方兴未艾，拉丁美洲的摩天大楼却相对较少。目前，墨西哥城已着手挖掘独立200周年庆典的经济潜力和纪念价值，准备建造新的摩天大楼——Bicentenario塔。

　　沿改革大道呈链式分布着多栋高层建筑，一直延伸至最大的城市公园——查普尔特佩克公园。Bicentenario塔选址就在该公园的东北角，毗邻两条城市主干路。Bicentenario塔的建造将会使这条建筑链延长至公园周围，构筑新的城市地标，带动并影响公园的边缘空间、主要基础设施建设和整个城市发展。Bicentenario塔为墨西哥城提供了急需的3A写字楼，同时还能为周边社区和公园游客提供良好的公共设施——空中大厅、会议中心、商店和餐馆。由两个金字塔堆砌起来的形状使整栋建筑看似熟悉却又新颖，既具历史意蕴又有超现实成分。

　　Bicentenario塔所采用的造型使它与周边环境产生动态联系，其中的公共空间均在两个"金字塔"相接的地方。两个金字塔在标高100 m处相接（这也是周围建筑的水平基点），组成通往公园的高空之门。Bicentenario塔凸起部分朝向公园和改革大道，连接历史悠久的城市中心轴线，而两个侧面在朝向公园的同时面对Los Lomas地区伸展开来，塑造出整栋建筑庄严、肃穆的形象。

　　摩天楼往往使自身特点内在化，Bicentenario塔中心大堂就凸显这一优势，将自身"隐蔽"在引人注目的空间中。在塔的最宽处设有一段贯穿空间，这样既便于自然采光和通风，同时也使内部空间不再孤立。中心大堂内层表皮的50%采用反光玻璃板，这种设计确保光线穿透程度的最大化。大堂空间在中部弯折，底部向公园开敞，顶部向城市开敞，这样既改善大楼的孤立状态，又使其有机地融入周围环境。Bicentenario塔正立面是组合结构的幕墙系统。薄薄的构件遵循大楼外形轮廓，构成由绳索组成的屏幕，对内保证大楼的透明度，对外则获得精巧的质感。垂直向上的幕墙在两个金字塔相接处被"压缩"，使公共空间最集中的楼层更趋于透明。

1 建筑与环境的融合

可持续方法：可以预见，Bicentenario塔将成为未来建筑的典范，因其能充分利用可持续和综合设计方法发挥建筑的最大效能。这就意味着，整个建筑系统会尽量地减少能源、水和建材的用量，最终目标是要减少使用过程中的碳足迹（二氧化碳消耗量对环境影响的指标）的数量以及建筑材料所耗费的能源。

气候：设计考虑到当地气候和环境，以优化能源使用、改善室内空气质量为目标。墨西哥城全年平均气温较为适中，大约17 ℃。冬天夜间气温能降至2.5 ℃，夏季白天最高气温达30 ℃。温度较高的几个月（大约从五月到九月）最为潮湿，白天平均相对湿度达到60%。冬季则较干燥。

空气和水：空气调节装置安装在大楼中央的设备层。这种设计增加地面净使用面积，而且集中式设备可以尽量减少维护、维修的费用，减少对写字楼租户的干扰。设备层的铺设同时也包括消防和生活用水设备。节水系统通过安装节水设备，处理和回收中水和雨水用于冲厕装置和园林灌溉，能节约30%的水量。

能量与光：高效率冷却机和锅炉设备安装在楼层较低的机械室内。冷却机和水泵上安装的变速驱动器受楼内能源管理系统控制，可调节能量的输入，从而准确地保证楼内的冷、暖需求。所有设备在设计时均留有一定余量，有效地保证在维护和维修时不干扰楼内设施的正常运行。楼内引入高压电（23kV）会通往低层电器室的高压开关，以供动力设备使用。

In 1810, Mexico gained its independence from Spain. 100 years later the Mexican Revolution began Mexico's political modernity. A century later Mexico and its capital stand at the brink of another quantum leap: a nation at home in a globalized world, in which economic prosperity and a new cultural flourishing promise to transform the nation and its capital.

Compared with the world's other economically ascendant regions such as Asia and the Middle East, Latin America has a skyscraper deficit. Poised to harness the economic and symbolic potential of the

Bicentennial, Mexico City will celebrate a historic moment with the emergence of a new skyscraper: Torre Bicentenario.

A chain of high-rises runs along the Reforma and continues around the city's largest urban park: Chapultepec. The site of the Torre Bicentenario lies at the northeast corner of Chapultepec Park, adjacent to the interchange of two major highways. The Torre Bicentenario will extend this line of buildings around the park, creating a new icon for the city. Located at the edge the park, major infrastructure and the city, the project has the potential to benefit all three. Torre Bicentenario will provide much needed class AAA office space for Mexico City, together with public amenities – a sky lobby, convention center, shops and restaurants – for the surrounding communities and visitors to Chapultepec Park. The stacking of two pyramidal forms produces a building simultaneously familiar and unexpected, historic yet visionary.

The form of the Torre Bicentenario creates a dynamic relationship between the building and its surroundings. Public programs are located where the two pyramids meet. The junction of the two pyramids occurs at 100m, the datum of the buildings that surround it, creating a dramatic new gateway to the park. The building bulges toward the park and the historic city center along the axis of the Reforma. While two sides stretch in the direction of the park, towards Los Lomas, the building is respectful and sober.

Skyscrapers tend to internalize their features. Atriums typically create dramatic spaces within, hidden from the city around them. Here, a void cuts through the building's widest point, providing access to light and natural ventilation and creating a relationship between the floors within. A pattern of reflective glass panels covering 50% of the interior surface maximizes light penetration. The void twists at its midpoint, opening at the bottom toward the park and at the top toward the city. Rather than exacerbating the skyscraper's isolation, it connects the building to its surroundings. The facade is a structural curtain wall system. The thin members follow the building's form creating a screen of cables that simultaneously maintain the building's transparency from within and create a sense of refined solidity when seen from a distance. At the point where the two pyramids meet, the facade is compressed, providing greater transparency at the building's most public levels.

Sustainable Approach: It is envisioned that the Torre Bicentenario will be a building of the future and an example of how a sustainable and integrated design approach can achieve maximum building performance. To this end, the design of the building systems will aggressively target the reduction in energy, water use, and the constructed volume. The ultimate goal is to reduce the building's carbon footprint in its operation and the embodied energy used in the materials to construct it.

Climate: The building systems shall be designed to respond to the local climate and environment to optimize energy use and improve indoor air quality. The average annual temperature of the city is mild, at about 17 ºC. Winter temperatures may fall to as low as 2.5ºC at night while, in the summer, high daytime temperatures may reach 30ºC. The warmer months roughly from May through September are also the most humid. The average relative humidity during daytime at these months is 60%. Drier conditions occur during the winter months.

Air and Water: The building will be served by air handling units located on central mechanical floors. This approach maximizes the net usable floor area and also centralizes equipment to Minimize maintenance cost and the disruption to office tenants when systems are serviced. The mechanical plant floors shall also be used to house the fire protection and domestic water equipment. Water conservation strategies shall be developed to reduce building water demand by 30%, by installing water efficient fixtures and capturing, treating and reusing gray water and rain water for flushing toilet fixtures and landscape irrigation.

Energy and Light: A high efficiency chiller and boiler plant shall be installed in the lower level mechanical room. Variable speed drives shall be provided on chillers and pumps and controlled by the building energy management system to adjust the energy input to exactly match building cooling heating demand. All equipment shall be designed with redundancy to allow for effective maintenance and repair without disrupting the operation of the building. The incoming electrical service will be high voltage 23kV, and terminated onto main high voltage switchgear located in a lower level electrical room dedicated for the utility equipment.

2 夜景效果

HIGH-RISE BUILDING | DESIGN WORKS

3　建筑形体模型
4　趋于透明的幕墙系统

高层建筑 | 设计作品

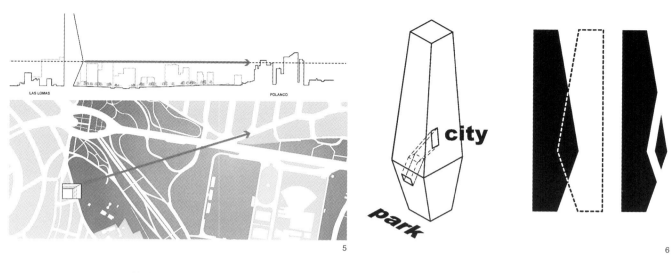

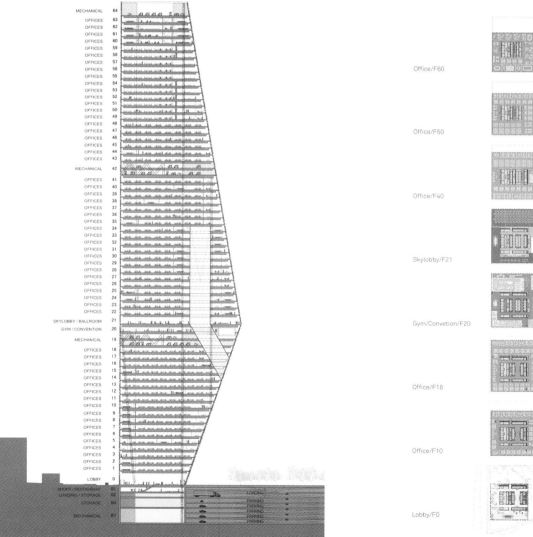

5　视线分析
6　公共空间与体量构成关系
7　剖面
8　平面

175

TIANJIN SINOSTEEL INTERNATIONAL PLAZA
天津中钢国际广场

MAD建筑事务所 | MAD Ltd.

项目业主： 中钢国际广场（天津）有限公司
项目类型： 办公楼、酒店、高档公寓、商业综合体
建筑面积： Tower A 228 129 m^2；Tower B 64 176 m^2
建筑高度： Tower A 358 m；Tower B 95 m
项目建筑师： 马岩松，党群
设计团队： 刘小普，Eric Spencer，Tony Yam，So Sugita，赵伟，李洁苒，吴非，
向明，周贞徽，Dominika Placek
合作工程师： 上海江欢成建筑设计有限公司
结构工程师： 中建国际
设备工程师： 诚中国
幕墙顾问： 迈进外墙建筑设计咨询有限公司
项目管理： 北京中科国金工程管理咨询有限公司
交通顾问： 奥雅纳工程顾问

中钢国际广场位于天津滨海新区，是中钢集团辐射华北、西北和东北三个区域的运营中心、物流配送中心和科技研发中心，也是中钢在华北地区的总部。该项目的建成将为中钢集团提供国际化、专业化的发展平台，使其形成背靠北京、南依香港、服务内地、覆盖全球的发展格局。

中钢集团委托MAD为他们在该区域的一号地块设计一座高358 m的办公楼和一座95 m高的宾馆，并成为天津未来具有中国特征的城市标志物。从城市上空俯瞰中钢国际广场，它采用简洁、恰当的对角式布局，为城市赢得两块公共景观空间：首先，形成了滨江城市多媒体喷泉广场；其次，则是城市内部的绿色森林花园。这种整体布局方式，既为办公楼和酒店营造了优质的生态环境，又为公众提供了优雅的休闲场所。

进入中钢国际广场的内部，在办公空间设计中，其标准层平面由边长50 m的正方形构成，进深12 m，楼层内不设柱子，可灵活满足客户的不同需求，实现小开间的分隔布局。办公区标准层高为4.3 m，办公室净高3 m，符合国际A级写字楼标准。广场内设30部高速电梯，两个大厅作为转换层，使用者可快速、便捷地到达各办公区域。因而，办公空间通过竖向交通的连接，贯穿起不同的水平层面，构筑出整体、开敞和自由的空间意象。

中钢国际广场立面采用"六棱窗"作为母题，这种造型取意于中国古典园林建筑。六棱窗不但成为独特的视觉因素，更成为独特的造型语言，构筑出崭新的、富有传统意蕴的现代建筑形象。六棱窗的层层叠加使得整栋大厦就如同是由无数生命细胞构成的、在持续不断生长的有机整体。

在这里，六棱窗除去造型作用，还是支撑这栋超高层的主要结构系统。六边形的不断叠加使高层外筒自身就具备明显高于普通梁柱结构的抗侧力稳定性，使整栋大厦不再需要柱子，外墙本身就是结构，实现办公标准层空间使用效率的最大化，同时减少了建筑的用钢量，节省了结构建造成本（这种结构已经通过全国超高层结构专家的可行性论证，这在世界高层建筑领域也属领先结构体系）。

大厦外立面的开窗不再采用传统玻璃幕墙，而是结合美国LEED环保节能建筑标准，针对天津各季节的风向和日照，调整建筑各个部位的开窗大小，提炼最合理的窗地面积比例，最大限度地减小热损失，实现节能和生态的目标。

综上所述，中钢国际广场以其独特的规划布局、先进的设计理念标志着中钢未来的发展姿态。同时，它蕴含传统文脉的造型意象，将中国古典元素与现代语汇相结合，赋予高层建筑一种区域归属感，在创新的同时，也标志着基于传统的开拓力量。

1 临水效果

2　入口空间
3　六棱窗的造型

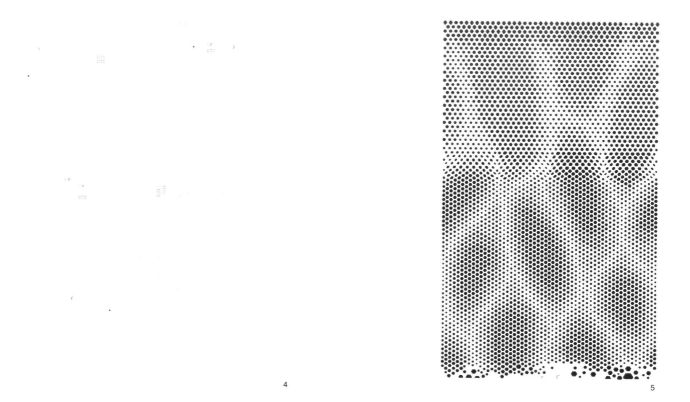

4
5

4 总平面
5 表皮的六棱窗主题示意
6 表皮夜景效果

6

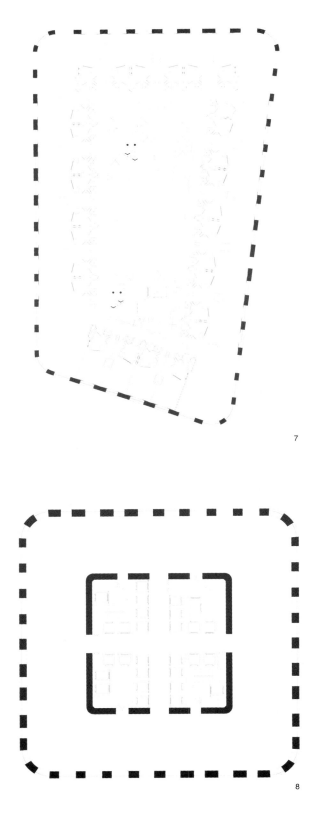
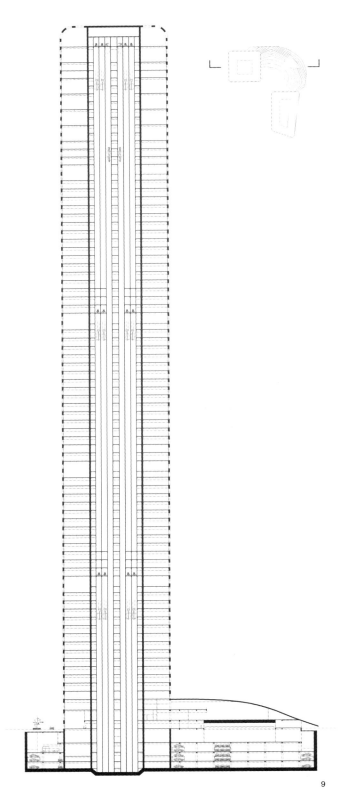

7 酒店层平面
8 办公层平面
9 剖面

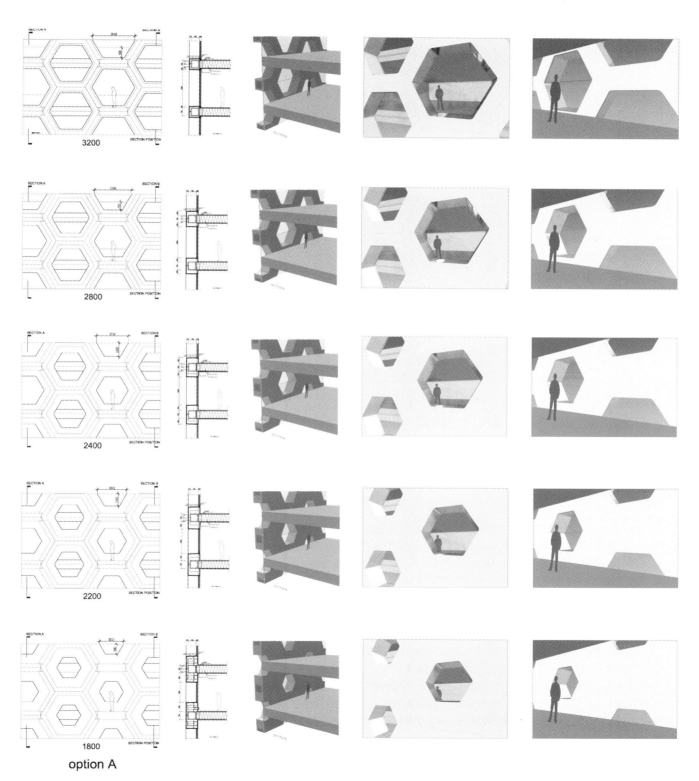

option A

10 六棱窗的造型分析

TORONTO ABSOLUTE TOWERS, CANADA
加拿大多伦多梦露大厦

MAD建筑事务所 | MAD Ltd.

项目业主：Fernbrook Homes/Cityzen Development Group
项目类型：公寓住宅
建筑面积：塔楼A 45 000 m²；塔楼B 40 000 m²
建筑高度：塔楼A——56层，170 m；塔楼B——50层，150 m
总建筑师：马岩松、早野洋介、党群
设计团队：沈军、Robert Groessinger、Florian Pucher、易文真、郝奕、要梦瑶、赵凡、柳苑、赵伟、李娟琨、于魁、Max Lonnqvist、Eric Spencer
合作建筑师：BURKA Architects INC.
结构工程师：SIGMUND，SOUDACK & ASSOCIATES INC.
设备工程师：ECE Group
电气工程师：ECE Group
景观建筑师：NAK Design
室内设计师：ESQAPE Design

"建筑是居住的机器"。但是我们必须发问，在"机器"及其所代表的社会基础发生如此剧变的今天，我们应怎样理解建筑？如果建筑正远离工业化，那么它更应该表达怎样的内容？

与北美其他迅速发展的近郊城市一样，加拿大多伦多地区的密西沙加市也一直在努力寻找自身的性格和定位。我们认为这样一座迅速发展的城市蕴藏着新的机会，它没有必要再像那些典型的城郊一样，一心梦想变成大都市，并有可能反思自身独特的地域性，考虑用一种特殊的方式回应日益膨胀的城市需求。

我们的设计不再桎梏于现代主义的简化原则，而是表达出一种更高层次的复杂性，更多元地接近当代社会和生活的多样化，满足多层模糊需求。我们认为建筑本身是对自然和环境的反映和强调，高层建筑设计一直是对地理因素和社会环境的一种强有力的陈述。

The Absolute Tower位于密西沙加市最重要的Hurontario街和Burmhamthope路交汇处，其所具有的重要性和标志性将使这片区域成为这个低密度近郊城市的中心。在我们的设计中，连续的水平阳台环绕着整栋建筑，而传统高层建筑中用来强调高度的垂直线条被取消了，整栋建筑在不同高度进行着不同角度的逆转，以呼应不同高度的景观文脉。

我们希望The Absolute Tower可以唤醒大城市中人们对自然的憧憬，感受阳光和风对人们生活的影响。The Absolute Tower也被当地民众和媒体称为"玛丽莲·梦露大厦"。

1 全景鸟瞰

HIGH-RISE BUILDING | DESIGN WORKS

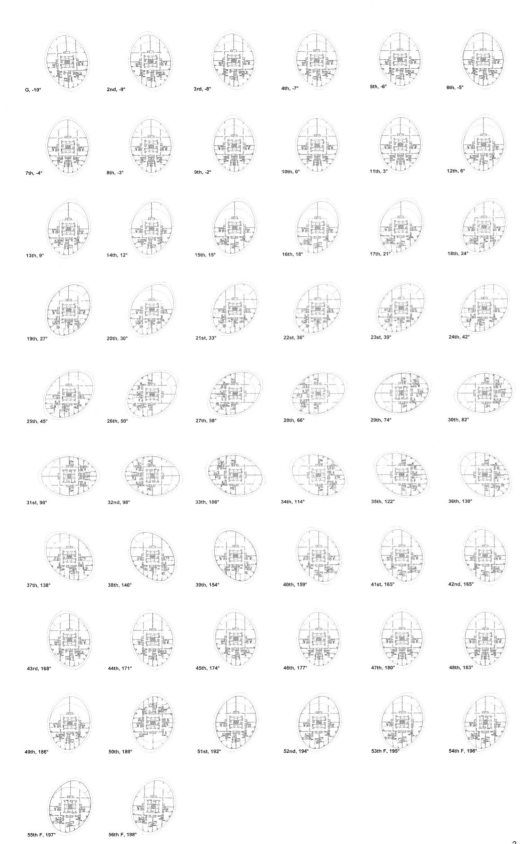

2 各层平面

3 整体效果

TRADE FAIR CORPORATE HEADQUARTERS, ITALY
意大利商品交易会公司总部大楼

意大利IaN⁺建筑设计 | IaN⁺

项目名称：商品交易会公司总部大楼
业　　主：Balticgruppen AB
建设地点：Kungsgatan, Vmeå, Sweden
设计单位：Henning Larsen 建筑师事务所
建筑面积：6 000 m²
建筑层数：7 层

建筑材料：松木镶板，玻璃幕墙
建筑设计：White Arkitekter, Tyréns, WSP Group, TM-Konsult
设计时间：2009 年
建成时间：2011 年
图纸版权：Henning Larsen 建筑师事务所
摄　　影：Åke E:son Lindman

会场内的工程

米兰商品交易会公司总部大楼设计方案源自一种空间理念而非一项造型创作，它重点关注工程的相互联系和功能特质，使设计独具美学特色。工程力图同福克萨斯（Fuksas）设计的各展馆直接相连，构成中心轴线，与多米尼克·佩罗（Dominique Perrault）楼进行远程的连接，令建筑兼具横纵两个方向上的张力，并与周围的"建筑公园"融合起来。

点和线编织出整个交易会场地，而建筑与建筑、建筑与周边环境相互交织，构成了独特的会场景观。Fuksas馆诠释了地面建筑的功能及造型，而包括酒店塔楼在内的Perrault楼则为园区的高空轮廓增色不少。方案通过"静肺"绿化区、天空与商品交易会场内各建筑之间的内在联系，着力强调总部大楼各部分间的视觉连接，不但强化了"建筑公园"这一概念，也加强了建筑同景观的衔接。

设计方案的选择

对设计方案起着决定性作用的条件可归纳如下：

第一，基于构建东门道路的愿景，以令现有建筑特色更为突出；第二，建筑新颖独特的体量、空间设计及其与其他建筑的遥相呼应，共同实现了建筑与整个会场体系的完美协同；第三，经过审慎的调查研究及对已有写字楼各立面的解构，设计师打破了水平滑动窗体立面和连续立面的三维外观；第四，基于上述方案的选取，设计内部空间及结构，营造品质及舒适度更高的办公空间。

现有的会场建筑是一块水平伸展的巨大空间，来自四面八方的人流会聚于此，是一处完美融入景观之中的地标。我方的设计具有与此相同的特色：它同样是一块巨大的空间，但建筑向竖直方向延展，突出了两个对立体量间的紧密联系。

根据功能及组织结构的划分，新建筑由不同的两部分对立构成，从而创造出动态的空间穿插造型。最终，生动的中央空间俨然成为了第三座楼体，令会场设计的经典元素在新建筑中得以再现，起到同会场以及周围景观之间建立空间与视觉联系的作用。办公室中的职场精英们恰似交易会会场充满生机活力的游客们，赋予全新的展会总部大楼活力。中间穿插的空间是一座真正的空中花园，它分置于几个全景露台之上，连接着工作区及其周边的所有自然、人工景观，供展会集团公司内各企业员工社交及休息之用。

建筑立面

总部大楼的构思意在突出城市元素，力求摒弃办公楼的传统外观，为新建筑营造一个富有个性的建筑形象，同时与2015年米兰世博园建筑风格整齐划一。

虽然如上所述，方案力图再现20世纪米兰的传统建筑风格，但它并同过去20年已变得尤为紧迫的可持续性问题及相关要求发生冲突。新建筑力求颠覆典型的玻璃写字楼设计传统，为其增添浓厚的城市气息，更重要的是这有助于将设计精力集中于立面系统及建材的选用，从而控制并降低能源消耗，增强建筑的生态可持续性。除此之外，立面原型设计中不透明板材交错排列，逐步淡化玻璃颜色，突出表现中部空间。设计师Massimiliano Fuksas运用相同的手法，通过起伏的穹顶造型成功地去除了结构轴线。我们力图重新使用一种传统的不透明材料来达到相同的视觉效果，扁菱形板材造型弱化了建筑体量，突出了它的空间感。此外，我们创新性地应用传统不透明材料多孔的材质特点诠释了新总部大楼的建筑体量，而整座建筑也因此立面系统在可持续性方面跻身A类建筑之列。

室内空间设计

整个公共空间及工作区设计均以实现最舒适的室内环境为目标，并使用户最大限度地欣赏展会场地建筑全貌。

在各公司布展空间的分配与协调方面，我方尤其专注于灵活与调节的理念，根据不同的运营组织模式来分配各个空间。灵活性这一概念并非指无限的空间组合及利用，而是指选择并确定定量与变量控制下的不同空间布局方案。空间分割以及墙体有条不紊地按既定模式进行布置，保证办公区的理想尺寸，通过可移动墙壁实现了空间的完美协调分配。

不同楼层围绕中央区域组织展开，中央区域包含全部的垂直连接部件、轴设备以及地面技术功能模块，并将每层楼分为3个区域：边缘区域——设有集会区及休闲区的外围区域；中央区域——围绕建筑核心展开，员工及管理层办公室均设置于此，根据使用需要分为独立办公室及开放型办公空间；悬臂区域——朝向中央空中花园，因悬臂长短不同而尺寸各异，行政办公室、会议室等功能模块设置于此，位置得天独厚，可纵览中央空间这座"第三号建筑"的美好景致。

Our proposal for the Milan Trade Fair corporate headquarter departs from an idea of space rather than from an idea of shape, being mainly concerned about the relational and functional characteristics of the project that will shape our design and characterize its aesthetic quality. The project intends to establish a direct connection with Fuksas pavilions and connection axis, and a remote connection with Dominique Perrault building, creating a horizontality-verticality tension that merges the "architecture park" in its surroundings.

Lines and points organize the system of the fair, buildings that are not self-referential but dialogue with the surroundings structuring the landscape. Fuksas defines the horizon, Perrault, inclining the hotel towers, depicts the sky. The introduction of the Fair headquarter in our

1 建筑与周围环境的融合

2 临街效果
3 建筑细部

proposed configuration accentuates the visual connection among the various elements articulating the complex through the relationship established among the central green lung, the sky and the Trade Fair buildings, strengthening the concept of "architecture park" and its relationship with the landscape.

Architectural choices

The main choices that drove our design solutions can be summed up in the following points: 1.the desire to define and structure the access to the East Gate, emphasizing the characteristics of the existing buildings. 2. The design of the new building volumetric and spatial articulation that could define and highly visible and communicative element, in a perfect synergy with the Fair system. 3. A formal research intending to de-structure a given model of office facades, suggesting an apparently tridimensional facade breaking off the model of the horizontal sliding window strip or the continuous facade. 4. The determination to structure an internal spatiality and organization that, within the above mentioned choices, could gain high quality and comfort working spaces.

The existing Fair Trade is a big void stretching horizontally and gathering the tension of all the flows crossing it, a sign in the landscape that turns out to be landscape itself. The element characterizing our proposal is exactly the same: a big void, stretching and expanding vertically emphasizing the tension between the two facing volumes.

The new building, within its functional and organizational organicity, is composed by two different volumes one facing the other, so to define a dynamic interstitial space. The result if the activation of the central void, that becomes a sort of a third building, a precious component of the new system and of its spatial and visual relationship with the Fair and the landscape: in the same way visitors live and make the fair space active, the office employees activate the vertical core of the new fair headquarter. This interstitial space is a real vertical garden, organized on various panoramic terraces, connection places between the working area and the entire natural and artificial landscape around it. The relax and collective areas are mostly concentrated in these zones, defining a highly relational dense space, open to interaction and socialization among the employees of the various companies that are part of the Fair corporation.

The Facade system

The new headquarter towers were conceived as urban elements, providing the new buildings with a strong architectonic identity trying to get rid of the "managerial" look, and considering the Fair area as an already consolidated and structured fabric by the 2015 Expo area.

Our determination we referred above to, to reconnect to a certain 20th century Milan architectonic tradition, does not go beyond certain sustainability issues and requirements, that in the last twenty years have become always more urgent. The new project, intends to

4
5

break the typical office glass building tradition, in order to give a more urban character to the towers and, above all, to focus our attention on the use of systems and materials that can control and reduce the energetic consumption and promote the environmental sustainability. Furthermore, our choice to adopt a facade composed by fiber reinforced colored concrete panels is a choice of field, an attitude that does not want to work on the mimesis with the surroundings but is more interested to strengthen the idea of continuity also through the chromatic integration with the existing structures, the landscape and the sky. Our facade prototype alternates the opaque panels matericity to the glass lightness characterizing the central void. In the same way Massimiliano Fuksas achieved to dematerialize the structuring axis by way of the canopy sinuous shapes, we tried to reinvent a traditional opaque material to accomplish the same effects; the rhomboidal panel geometry creates a vibration dematerializing the volume and accentuating it spatial qualities. We did not design a typical office building facade, but we have rather reinterpreted the traditional use of opaque materials, their porosity and texture, to characterize the volume of the new headquarter building and gain a facade system that contribute to classify the entire building as A class from the sustainability point of view.

The interior space definition

All the communal spaces and the workspaces have been designed in order to maximize the internal environmental comfort and make the most of particular privileged views on the Fair structure.

On the definition and calibration of the spaces to allocate to the different companies we particularly focused our attention on the flexibility and adjustment concept, to the disposition of those spaces to the different working organizational models. The flexibility concept is not directed to the research of an infinite space combination and use possibilities, but rather toward the determination of different space layout solutions characterized by some invariants and variables. The space partition and the facade modulation is marked the rhythm of 1,30 mt module assuring an optimal range of office space sizes and the perfect space modulation through movable partitions.

The different floors are organized around a central core, containing all the vertical connections, the installation shafts and the floor technical functions. This central core divides the floor in three different areas: the exterior border areas where collective spaces and relax areas are located, the central area, around the core, where the employees and management offices are located, either in cubicle, offices of open-space configuration, according to the needs, the cantilevered area opening toward the central vertical garden, variable in dimension according to the cantilever length, and where the more representative functions are located, like executive offices, meeting rooms, offering a advantaged and beautiful views on the "third building", the central void structuring the complex.

4 立体绿化
5 建筑远景

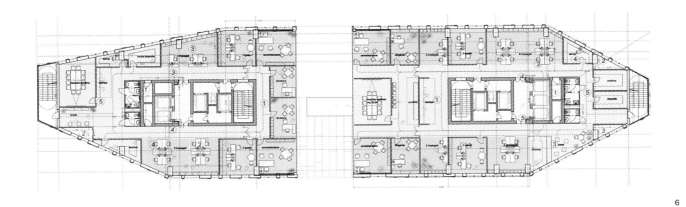

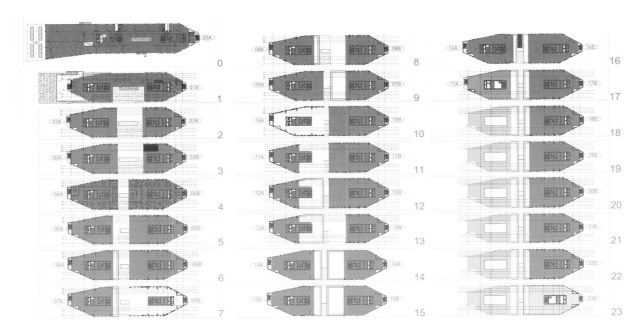

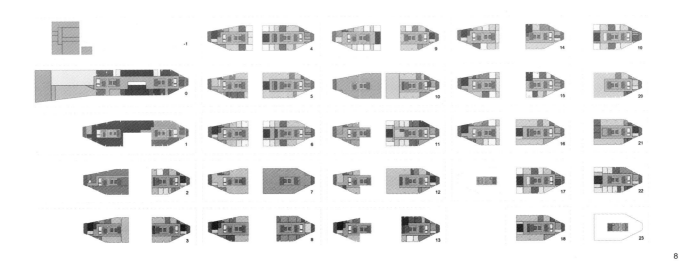

6 标准层平面
7 各层平面
8 功能分区

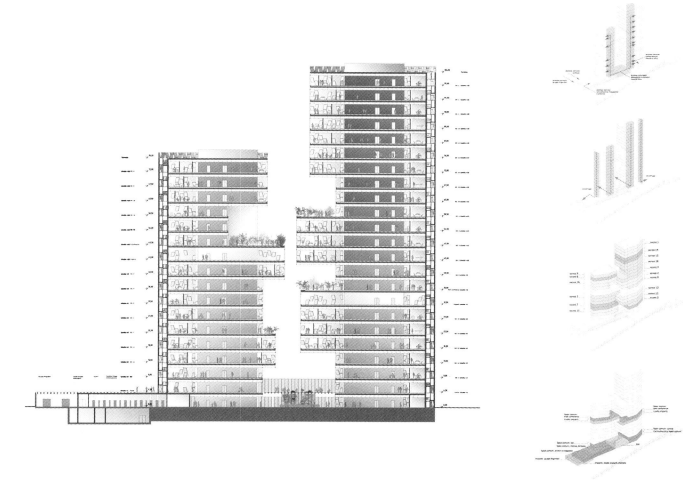

9 内部空间
10 剖面
11 功能分析

CHEONGNA CITY TOWER, KOREA
韩国青罗城市大厦

意大利IaN⁺建筑设计 | IaN⁺

建筑事务所：IaN⁺, Roma (Carmelo Baglivo, Luca Galofaro)
设计团队：Joshua Mackley, Andrea Leonardi, Agita Putnia
结构工程：Sistema Duemila Spa-Milano
设计团队：Alessandro Vigano', Stefano Reale, Valentina Zaletti di Arpi

塑造仁川的整体形象

设计方案旨在全力帮助仁川树立一个极具说服力的城市形象。拱形结构作为一种标志力图描绘出这一区域的生机与活力；拱是连接传统与未来的纽带，它动感的造型以及创新的结构无不代表了建设"东北亚新经济中心"的活跃思想以及乐观的愿景，它作为区域地标与国家地标是自由经济区的象征。

通过展示建筑基址及材料选择过程中为适应当地环境条件所采用的现代科技，青罗城市大厦（Cheongna City Tower）为区域经济及文化领域的发展做出了很大的贡献，令该地区成为汇聚科技、景观、投资及人力等资源的"特区"，从而使其拥有了人居生活、办公及娱乐等活动的理想环境，强化了其作为国际化城市的整体形象，使其在发展与创新方面引领世界。

从许多方面来说，拱的造型大多是东北亚各国商业及政治之都的标志性入口。全新的经济设施及标志性的青罗城市大厦促进了国外投资，从而推动这一新兴东北亚经济中心的区域性平衡发展。

这是一座全新的望高塔，通过卓越的垂直造型元素，与周边环境形成了一种强烈的反差效果。同时，它还吸引行人驻足，营造出天空轮廓以及肉眼所能观察到的独一无二的视觉体验。从功能层面而言，各种相关娱乐休闲设施赋予景观、游览者以及仁川市全新的意义。

整体规划

这个设计方案是建筑与自然的一次无缝融合，它找到了景观与建筑环境间的平衡点。望高塔位于水晶公园轴线与城市运河轴线交叉地带的核心，因而成为该区域的焦点所在。耸入云霄的望高塔兀立于公园中央，俨然成为城市新形象的闪光徽标之一，甚至可以说此方案是仁川乃至韩国最易辨认的标志之一。

高塔优雅的造型设计在多个层面得以体现：拱结构让人回想起历史上的经典拱门构造，而高塔的结构以及几何形态则除现代感以外，另有几分创造与未来的韵味；从城市以及运河的角度来看，拱结构如同一座敞开的大门，吸引游客入园，有助于盘活运河滨水区及城市的中轴线；自公园望去，高塔独一无二的造型，好似从公园中生长出来一般，建筑在公园中的选址及其在仁川市的重要地位，使得以上之解读瞬息可辨。

Our design aims to help create a potent expression of Incheon identity. The Arch as an icon seeks to portray the vibrant condition of this region: an arch as a connection between tradition and future, its dynamic form and structural innovations as a symbol of the optimism and the active spirit of "Northeast Asia's new dynamic center", a symbol of the free economic zone as regional and national landmark.

By showcasing modern technologies adapted to local environmental conditions in siting and materials, the Cheongna City Tower celebrates the region's dynamic economic and cultural expansion into a "special district" where technology, landscape, investment and manpower come together to have an ideal environment for living, business and leisure, strengthening its stature as an international city and giving it a place as a world leader in development and innovation.

The Arch becomes in many ways an iconic gateway for the business and political capitals of north-east Asian countries; the implemented conditions of the area, the new economic facilities and the new Cheongna City Tower as a symbol will foster foreign investment in order to create a regional balanced development as a new economic center of northeast Asia.

The new observation tower establishes a monumental dialogue with its surroundings through a prominent vertical element. At the

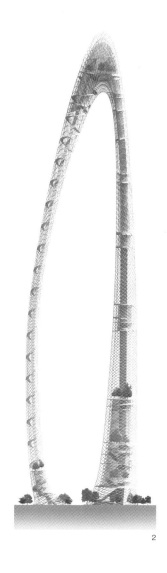

1 全景效果
2 立面图

same time it engages the visitor at the pedestrian level, to create a singular experience both as an element of the skyline and at a personal scale. In function, the creation of an observatory tower and related amusement-leisure facilities gives new meaning to the relationship between the landscape, visitors, and the City of Incheon.

Our proposal for the project site is a seamless union of architecture and nature, a balance of landscape and the built environment. The tower's placement at the pivotal intersection between the axis of the Crystal Park gardens and the axis of the urban canal gives it heirarchical importance as the focal point for all development on the site. Standing alone in the center of the park, the soaring tower becomes a shimmering icon for the new identity of the city. The tower's singularity allows it to be the most recognizable symbol in Incheon, and perhaps in all of Korea.

The graceful form of the tower also operate on multiple levels; the arch form recalls classical and historical use as a gateway and as a place of great importance, while the structure and geometries of the tower are not only contemporary but inventive and futuristic. The form of the arch also allows the tower to address its context in diverse, subtle ways. From the perspective of the city and the urban canal, the arch opens like a gateway beckoning the visitor to enter the park, helping to activate the waterfront of the canal and enliven the central urban axis. From the view of the park, the tower stands as a more monolithic, singular form, seeming to grow from the gardens and standing as a tree or reed in its pristine natural setting. The tower's placement in the gardens and at the moment of key importance in Cheogna allows all of these readings to be understood simultaneously.

HIGH-RISE BUILDING | DESIGN WORKS

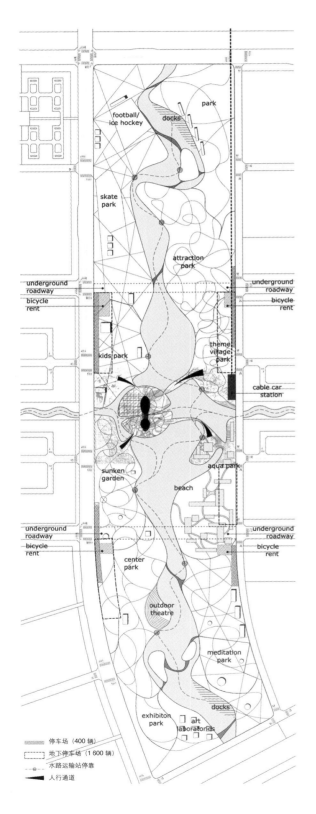

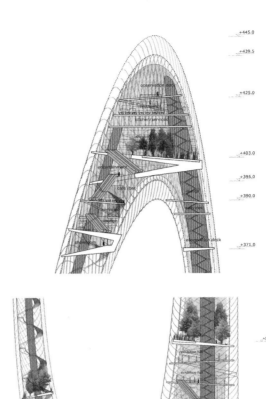

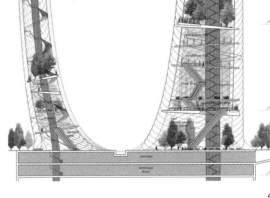

3 园区总平面
4 建筑顶部和底部剖面
5 总平面

6 仰望顶部
7 室内空间
8 入口效果

CPH ARCH, DK
丹麦巨拱

丹麦3XN建筑事务所 | 3XN

项目位置：哥本哈根 Marmormolen 地区
项目客户：By og Havn
项目规模：62 000 m²
建筑设计：丹麦 3XN 建筑事务所
项目摄影：Adam Mørk

桥梁是陆地景观中最为有力的建筑形式。它横跨深流的水体，连接起两片延伸出来的大陆，形成了唯一没有水的连接点。另一个典型元素就是城门，它既标志了乡村和城镇之间的边界，又从物理、结构和美学角度上包围着城镇。位于Marmormolen的建筑提案涉及到这两方面：该建筑既是将Marmormolen和Langeliniekaj两地连接起来的桥梁，又是这里的城门。这个建筑为哥本哈根港湾营造了统一的区域。设计理念就是创造出一个将独一无二的多样化发展中的复杂城市环境与可能建成的灵活有效的商业区连接在一起的结构。塔楼、桥梁和其他建筑元素构成了一个单一、漂浮、动态的趋向。从方案设计和外表面两个设计角度来说，显眼的横跨港口的跨桥则是这一趋向的独特所在。横跨港湾的拱桥的建立彻底地改善了公共的出入情况，并为该地区的生活和发展创造了崭新的机会。

从建筑和发展的角度来说，在入港口的两侧有着朝向水流的趋向，这一趋向由都市的交通形式和普遍的公众流向所加强。这一趋势延伸至横跨港湾的拱门，创造出一个连续的层级关系，从而连接起位于港湾两侧的大楼。同时，较高的塔楼被视作位于Langeliniekaj的现有建筑中一个自然的延伸，在另一侧较矮的塔楼则将与Marmormolen未来的发展相连接：一侧有所要求的蜿蜒而立的防噪音幕墙；朝向港湾一侧是更为开阔的建筑，在这里能够尽览美景。在Marmormolen一侧，行人们将从位于塔楼中心的中庭涌到坡度较缓的拱桥上。在Langelinie一侧，如果你想从跨桥到建筑外的海湾，可能需要您搭乘电梯然后坐到楼下的门厅。设计中的关键元素在于确保行人和骑自行车者经这个公共连接，并且穿过港湾时能够体验到连续的路线，以及位于两侧易于出入且无间断的道路。

塔楼和拱桥形成的这一门户式建筑无疑将成为这一地区、这一港湾乃至整个哥本哈根的一个新的地标式建筑。

这个港湾拱门的设计采用了三个简单的形状。朝西，在Marmormolen一侧是一个低矮宽阔的结构。朝东，在Langeliniekaj一侧是高耸且较为狭窄的大楼。而在两个塔楼之间则是横跨港口的典型拱桥。将最高的塔楼放在Langeliniekaj一侧实现了很多目标，而这一结果将有利于促进该地区的整体发展。将拱桥设置在码头尽头的三个最重要原因是光线，尺寸和可视性。将塔楼设在尽可能远的东侧，使塔楼仅在有限的时间内对开发区域造成阴影。而且，从高度方面来说，这样的位置安排从最高的塔楼到横跨港湾的拱桥再到矮一些的塔楼之间形成了一个自然的层级关系，并与未来建设的联合国村的高度形成和谐的关系。建筑的高度随着趋向降低，这使得港湾两侧也成为这一自然流向的一部分。与此同时，这一自然的流向与作为地标和灯塔的较高一侧的塔楼形成对比。

A bridge spanning a body of deep water, providing the only dry connection between two stretches of land, is one of the most powerful architectural experiences in the landscape. Another classical element is the town gate, which marks the boundary between the countryside and the town, and "contains" the town, physically, structurally and aesthetically. Our proposal for a construction on Marmormolen is both: a city gate and a bridge that links Marmormolen with Langeliniekaj, creating a new coherent area in the Port of Copenhagen. The idea is to create a structure which brings together a complex urban situation in a distinctive and diverse development with the possibility of including flexible and efficient business areas. The towers, bridge and the other building elements constitute one single, floating dynamic movement, characterised by the bold span across the harbour entrance in terms of

1 滨水效果

both the plan design and the facade. Establishing a connection across the harbor radically improves public access and creates brand new opportunities for life and growth in the area.

On both sides of the harbour entrance there is, in terms of both architecture and development, a movement towards the water, which is reinforced by the urban traffic patterns and the public flow in general. Extending this movement in an arch across the harbour creates a progression that connects the buildings on either side of the harbour. While the tall tower will be perceived as a natural extension of the existing buildings on Langeliniekaj, the lower tower on the opposite side will be bound to the coming development on Marmormolen: in towards? sterbro in a snake that forms the desired noise screen; and towards the water in a more open construction which opens up for views. On the Marmormolen side, pedestrians will emerge onto the gently arching bridge from the tower's central atrium. On the Langelinie side, when you step ashore, so to speak, from the bridge, it will be possible to take a lift down to the foyer. A key element in the design has been to ensure that pedestrians and cyclists will experience the public link across the harbour as a continuous route and as an easily accessible continuation of the infrastructure on both sides.

The gateway formed by the towers and the bridge will unquestionably become a new landmark for the area, the harbour and all of Copenhagen.

The harbour gate has been designed using three simple shapes. Towards the west, on Marmormolen, a low, broad structure. Towards the east on Langeliniekaj, a taller, more slender building. And between the two, a classic arched bridge across the harbour entrance. Siting the tallest tower on Langeliniekaj achieves several things which will help to enhance the overall development of the area. The three most important reasons for positioning the bridge at the end of the pier are light, scale and visibility. With the tower placed furthest to the east, it casts a shadow on the developed area for limited periods. Moreover, this way there will be a natural progression in terms of scale from the tallest tower via the bridge across the harbour to the lower tower, in harmony with the scale of the future UN village. The scale declines in a continuous movement, which allows the two sides of the harbour to become part of the same organic flow. At the same time, the organic flow is contrasted by the tall tower as a landmark and beacon.

HIGH-RISE BUILDING | DESIGN WORKS

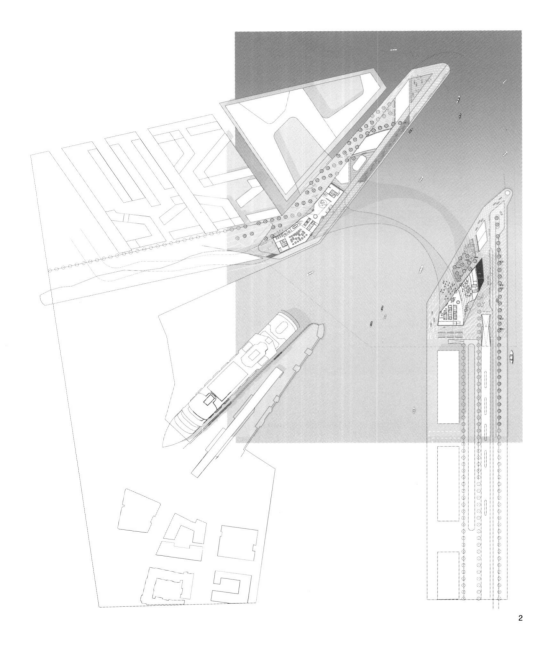

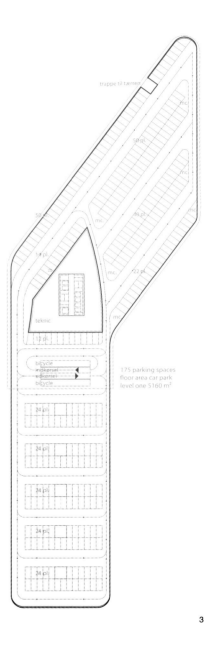

2 总平面
3 一层平面

4 临水立面

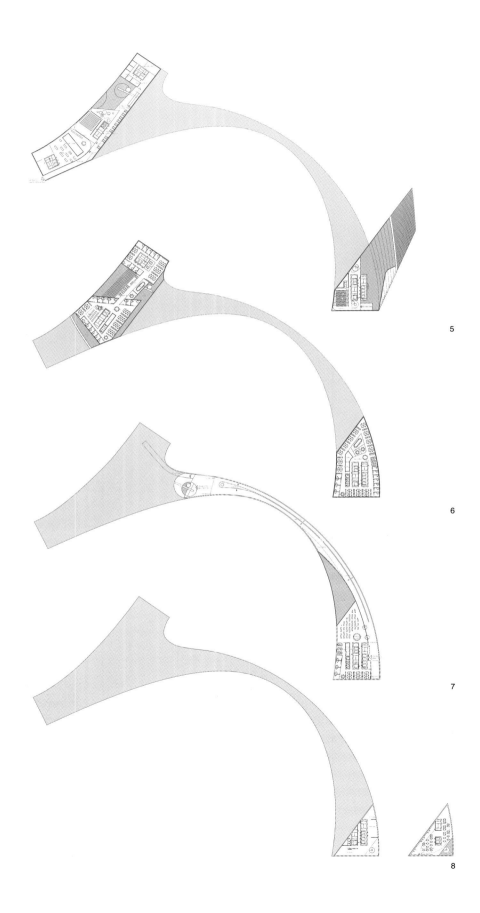

5 二层平面
6 十二层平面
7 二十二层平面
8 三十七层平面

9 与环境的融合
10 拱桥结构

NIJMEGEN BUSINESS AND INNOVATION CENTRE FIFTY TWO DEGREES, THE NETHERLANDS
荷兰奈梅亨商务及创新中心 52°

Mecanoo建筑事务所 | Mecanoo

方　　案：包含办公室、会议厅、1个剧场、餐饮娱乐设施、公寓、零售和运动设施的7万 m² 的多功能综合体
设　　计：2004~2005年，2002年竞赛一等奖
实　　施：2005~2008年
委 托 方：Ballast Nedam Bouw, Arnhem; ICE Ontwikkeling, Nijmegen
建 筑 师：Mecanoo建筑事务所，代尔福特 Mecanoo architecten, Delft
结构工程师：ARUP, Amsterdam (competition)；Adviesbureau Tielemans BV, Eindhoven
总承包商：Ballast Nedam Speciale Projecten, Utrecht
建筑服务顾问：Royal Haskoning, Nijmegen
建筑含服务造价：一期工程 4 200 万欧元
摄 影 师：Christian Richters

知识中心

新千年伊始，飞利浦半导体公司（现NXP）希望通过建造一个新的为电子工业服务的半导体研发知识中心来扩展他们在奈梅亨Nijmegen的生产和研究基地。现存的生产基地出于安全原因由栅栏封闭，而新的知识中心集技术、科学、文化、工作、生活和休闲于一体，增加不同方面的交叉和合作机会。该项目名称"52°"源自基地的维度，也寓意公司创始人的雄心——跨越整个世界。

连接

52°是环绕Neerbosscheweg区域的一项大规模总体规划的一期工程。86 m高的塔楼立于一斜坡上，不安定而又刺激地探入附近的高佛特公园（Goffert Park）。二期工程地面层将呈拱形覆盖Neerbosscheweg区域，建立起城市、高佛特Goffert体育场和新的高佛特Goffert轻轨站间的直接联系。植草屋面下是600个车位的停车场、各种商业设施以及一个含商店和餐厅的购物广场。预计二期工程还将在综合体内加入会议室、剧院、酒店、运动设施和商店。

扭结

塔楼为17层，其中下部的8层偏离垂直方向10°，形成一种对城市的邀请。弯曲的形式由混凝土和钢的混合结构建造，当高度改变时每个中间楼层都可以与混凝土核心筒保持联系。为了缩短建造时间（每周建1层的速度）而采用了预制覆盖层，像素式的图案赋予立面抽象的外观。一部惹人注目的曲线宽大楼梯从广场（Esplanade）延伸到接待厅，覆盖着桃花心木的波浪墙面充当视觉联系要素，指明通过建筑的路线。办公和试验楼层可以进一步灵活划分。气候天花板系统的使用使各个工作场所的空间环境变得可调节，从而这些空间能够适应新产品开发机构的需要。

Knowledge centre

At the start of the new millennium, Philips Semiconductors (now NXP) wanted to expand its current production and research site in Nijmegen by creating a new knowledge centre for the development of semiconductors for the electronics industry. The existing production site is sealed off and fenced in for security reasons. The new knowledge centre, where technology, science, culture, work, living and leisure come together, fosters chance encounters and collaboration with diverse parties. The name of the complex, fifty two Degrees, refers to the site's 52nd degree of latitude, which like the ambition of the initiators, spans the entire world.

Link

FiftyTwoDegrees is the realisation of the first phase of a large-scale master plan surrounding the Neerbosscheweg. The 86 metre tall tower stands on a slope and is fluidly yet excitingly absorbed into surrounding Goffert Park. The second phase will see the ground level overarch the Neerbosscheweg, creating a direct link with the city, the Goffert stadium and the new Goffert light rail station. Under the grassed roof are parking spaces for six hundred cars, various commercial facilities and a covered Plaza with shops and restaurants. Conference rooms, a theatre, a hotel, sports facilities and shops are due to be added to the complex in the second phase.

Kink

The tower is seventeen storeys high. The lower eight floors are ten degrees out of plumb, creating an inviting gesture towards the city. The bent form was created by the hybrid construction of concrete and steel, whereby for each intermediate floor the elevation shifts in relation to the concrete cores. In order to shorten the construction time - one floor per week - it was decided to use prefabricated cladding, resulting in a pattern of pixels that gives the facade an abstract appearance. A broad staircase leads from the Esplanade to the reception hall with its conspicuously curved wooden benches. The undulating wall clad with mahogany represents a visually connecting element that automatically indicates the route through the building. The office and laboratory floors can be flexibly subdivided. The use of climate ceilings allows the climate to be regulated for each work station so that the spaces can adapt to the organisation of new products being developed.

1 建筑实景

HIGH-RISE BUILDING | DESIGN WORKS

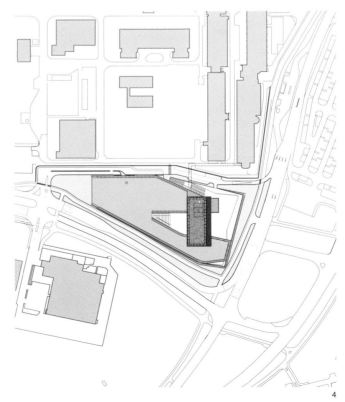

2　夜景效果
3　临街立面
4　总平面

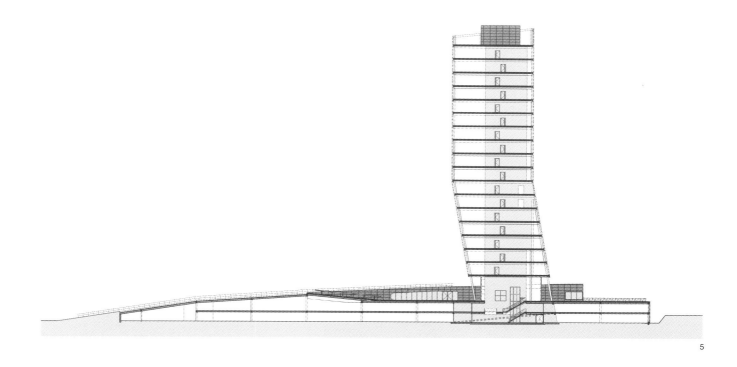

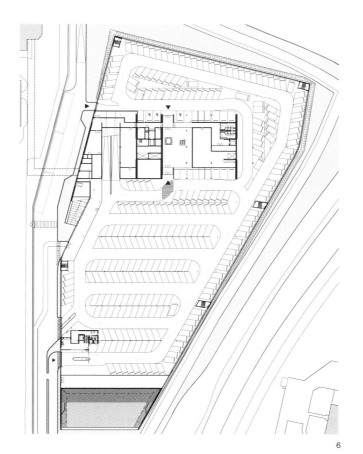

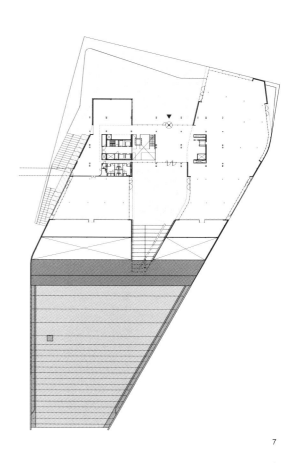

5 剖面
6 地下二层平面
7 一层平面

8
8 入口空间
9 室内空间
10 地下空间
11 入口前厅

9

10

11

HEARST TOWER, USA
美国赫斯特大厦

福斯特及合伙人事务所 | Foster + Partners

项目业主：Hearst Corporation
建筑设计：Norman Foster, Brandon Haw, Mike Jelliffe, Michael Wurzel, Peter Han, David Nelson, Gerard Evenden, Bob Atwal, John Ball, Nick Baker, Una Barac, Morgan Fleming, Michaela Koster, Chris Lepine, Martina Meluzzi, Julius Streifeneder, Gonzalo Surroca
设备配合：Norman Foster, Brandon Haw, Mike Jelliffe, Chris West, John Small, Ingrid Solken, Michael Wurzel, Peter Han
联合设计：Adamson Associates, Gensler
项目经理：Tishman Speyer Properties
建设高度：182 m/46 层

赫斯特出版公司的建造梦想源于1920 年，当时William Randolph Hearst 先生设想将哥伦布环岛变为娱乐和传播公司在曼哈顿的新中心。在第八大道上，赫斯特总部委托建造了一座6层高、极富装饰艺术风格的建筑，以此作为赫斯特在出版业的总部。1928年，当建筑完工时赫斯特又设想以这座建筑为基础，再建一座富有标志性的大厦，但这项计划却一直被搁浅。福斯特及合伙人事务所基于从德国新议会大厦和大英博物馆大厅中所获得的建造经验和国际声望，于70 年后接手设计建造这座大厦，其最大的挑战就是如何使新、旧建筑间产生和谐并具有创意的关系。

新建的42 层大厦是建在原有老楼基础上的，通过透明玻璃的包裹、贯通，使新、旧建筑和谐共融，大面积自然天光被引入建筑底层，新建塔楼仿佛凌空飘浮在底层之上。其中设计的突出亮点就是贯穿6 层的共享大厅。这个宏大的空间如同繁华的城市广场，其中包括电梯大堂、赫斯特自助餐厅、会议礼堂及为其他特殊功能服务的夹层空间。在结构选型上，新建大楼采用三角结构。这种高效的构筑方式对比传统框架结构可节约20%的钢材。在立面构图中，设计将三角形体块相交处向内翻转进去，创造了独特的多刻面效果，在勾勒轮廓线的同时也强调出建筑的垂直体量感。

大楼在生态方面的设计也十分出色。建造中再生钢材的使用量占总用钢量的85%，对比周围那些采用普通钢建造的建筑，可节约26%的能源。赫斯特大厦是该市新建筑，是最先获得美国绿色建筑协会能源和环保设计纲领 (LEED) 权威认证的设计。而作为一家公司，赫斯特总部非常重视健康工作的理念，认为宜人的工作环境将促进企业未来的发展。因而，赫斯特大厦的绿色建筑实践将引领城市中更多环保建筑的设计和建造。

The Hearst Headquarters revives a dream from the 1920s, when William Randolph Hearst envisaged Columbus Circle as a vibrant new quarter for media and entertainment companies in Manhattan. Hearst commissioned a six-storey Art Deco block on Eighth Avenue to house his publishing empire. When it was completed in 1928 he anticipated that the building would eventually form the base for a landmark tower, though no scheme was ever advanced. Echoing an approach developed in the Reichstag and the Great Court at the British Museum, the challenge in designing such a tower at some seventy years remove was to establish a creative dialogue between old and new.

The new forty-two-storey tower rises above the old building, linked on the outside by a transparent skirt of glazing that floods the spaces below with natural light and encourages an impression of the tower floating weightlessly above the base. The main spatial event is

1 纽约天际线间的赫斯特大厦
2 赫斯特大厦局部

a lobby that occupies the entire floor plate and rises up through six floors. Like a bustling town square, this dramatic space provides access to all parts of the building. It incorporates the main elevator lobby, the Hearst cafeteria and auditorium and mezzanine levels for meetings and special functions. Structurally, the tower has a triangulated form – a highly efficient solution that uses 20 percent less steel than a conventionally framed structure. With its corners peeled back between the diagonals it has the effect of emphasising the tower's vertical proportions and creating a distinctive facetted silhouette.

The new building is also distinctive in environmental terms. It is constructed using 85 percent recycled steel and designed to consume 26 percent less energy than its conventional neighbours. As a result, it is the first new occupied office building in the city to have been given a gold rating under the US Green Buildings Council's Leadership in Energy and Environmental Design (LEED) programme. As a company, Hearst places a high value on the concept of a healthy workplace – a factor that it believes will become increasingly important to its staff in the future. Indeed, Hearst's experience with the green building process may herald the more widespread construction of environmentally sensitive buildings in the city.

3 赫斯特大厦实景
4 赫斯特大厦剖面图
5 赫斯特大厦共享大厅平面
6 赫斯特大厦地上一层平面

高层建筑 | 设计作品

4

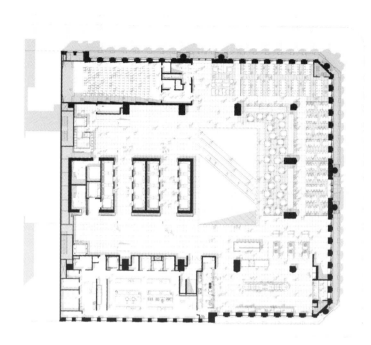

5

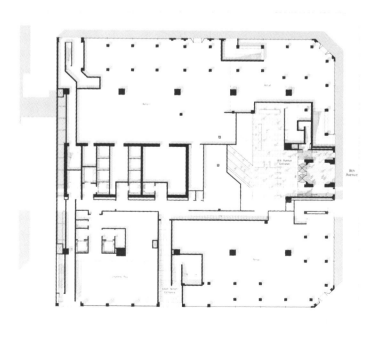

6

HIGH-RISE BUILDING | DESIGN WORKS

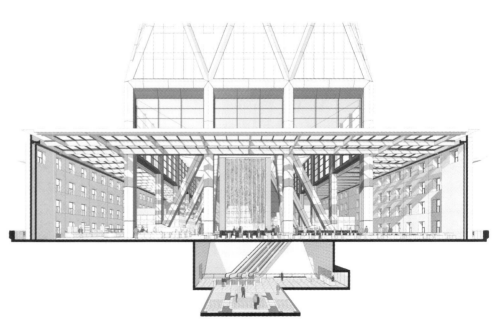

7 赫斯特大厦室内景观
8 赫斯特大厦室内景观
9 赫斯特大厦一层空间透视
10 从共享大厅仰望建筑顶部

TOWER TWO ON THE SITE OF THE WORLD TRADE CENTRE, USA
美国世界贸易中心二号塔

福斯特及合伙人事务所 | Foster + Partners

福斯特及合伙人事务所在格林尼治大街200号设计的一座78层办公大楼，是纽约世贸中心重建项目的一部分，它也是近年来城市规划和建筑设计领域最富挑战性的项目。福斯特的设计同时包含着纪念和再生的双重内容，其水晶般闪亮的建筑外观、钻石般璀璨的顶部造型为纽约天际线增添了新的景致。

办公大楼围绕中部的十字形核心进行布置，这种布局模式生成了四个建筑体块，灵活无柱并有自然光照的办公空间一直延续到59层，59层以上则以富有角度的倾斜造型与纪念公园产生空间呼应。格林尼治大街200号的独特区域位置赋予大楼独特的顶部创意，使其从任何位置看去都成为纪念公园所在地的象征性标志建筑。顶部楼层作为开敞的多功能厅，能俯瞰公园、河流和城市的壮观景象。

福斯特及合伙人事务所通过对大楼使用功能的深入挖掘，将格林尼治大街200号的街区结构、安全效能、环保策略和城市设计提升到新的层面。大楼的建造考虑了所在场地的几何形状，其十字形核心决定了结构骨架和布局模式。设计包括基本的垂直循环空间，高速电梯通往大楼中部的天空大厅，通过另外两组电梯连接更高的楼层。同时，由于每一楼层均设有准确方便的方向标，这种跨走廊式的横向循环系统连通了楼内的办公空间。

设计从十字中心向四周展开，恰当地将整栋大楼划分为四个相互联系的部分，在建筑立面上，通过凹槽强调了这种分割方式。建筑的功能布局由核心向周边展开，空间末端就会形成一些专属的灵活区域，这些区域可以是楼层间的楼梯或双倍高度的前庭。同时，它们也是大楼环保设计中不可或缺的部分，在春秋季节能将新鲜空气引入室内。格林尼治大街200号按照最高节能等级设计，将力争美国绿色建筑协会能源和环保设计纲领(LEED)黄金标准认证。

Foster and Partners has designed a 78-storey office tower at 200 Greenwich Street as part of the redevelopment of the World Trade Centre site in New York. One of the most important urban planning and architectural challenges of recent times, the concept is driven by memory, but equally by a sense of rebirth. Its sparkling glazed crystalline form and diamond shaped summit create a bold addition to the New York skyline.

Arranged around a central cruciform core, the tower comprises four blocks containing light filled, flexible, column free office floors that rise to the 59th floor, whereupon the glass facades are sheared off at an angle to address the Memorial Park. Giving the building its distinctive inclined summit, 200 Greenwich Street also acts as a symbolic marker of the location of the Memorial Park when viewed from any location. The upper floors contained within the summit provide the opportunity for spectacular multiple-height function rooms with sweeping views of the park, the river and the city.

A continuation of Foster and Partners' investigation into the nature of the tower, 200 Greenwich Street takes structural, functional, security, environmental and urban logic to a new dimension. The tower is informed by the geometry of the site, with the cruciform core providing the structural backbone as well as the key organising diagram. It accommodates the primary vertical circulation, with high speed shuttle elevators rising to an intermediate sky lobby where the upper floors are served by two further banks of elevators. It also allows for cross corridor circulation by providing excellent orientation at every level, and opening views out across the office spaces.

Extending the logic of the core, the volume of the tower is punctuated on all four sides by notches – elegantly breaking up the mass of the tower into four interconnected blocks. Towards the perimeter, the core culminates in dedicated flexible zones with the opportunity to create staircases between floors, and the possibility for double-height atria. These zones can be an integral part of the building's environmental strategy by drawing fresh air into the building during spring and autumn. Designed to the highest energy efficiency ratings, 200 Greenwich Street will seek to achieve the gold standard under the Leadership in Energy and Environmental Design (LEED) by the US Green Building Council.

1 世贸中心二号塔街景

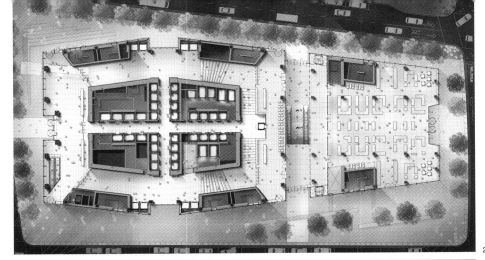

2

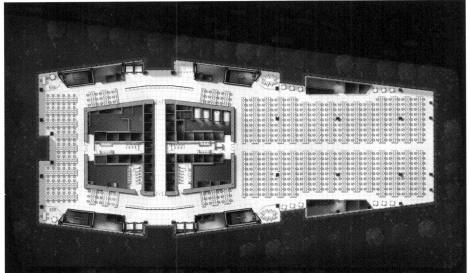

3

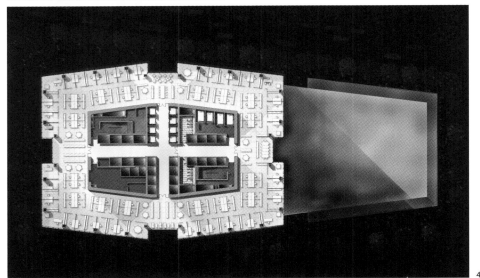

4

5

2 一层平面
3 商务层平面
4 办公层平面
5 办公层平面

6 临街效果

7
7 剖面
8 钻石喻意的顶部造型

ABU DHABI CAPITAL GATE, UNITED ARAB EMIRATES
阿拉伯联合酋长国阿布扎比资本中心塔

RMJM建筑设计集团 | RMJM Architects

项目业主：阿联酋阿布扎比国家会展公司（ADNEC Abu Dhabi National Exhibitions Company）
项目设计总监：Jeff Schofield
项目总建筑师：Tony Archibold
RMJM 设计团队（参与此项目的其他 RMJM 设计人员）：Greg Lau（RMJM 工程设计），Martin Wild，Anita Manoj，Andy Chen 以及 RMJM CAD 技术支持部门
建筑高度：161 m
项目规模：5 万 m²
项目造价：$8 billion
项目功能：商业、居住以及 Hyatt 酒店综合体

RMJM与阿布扎比国家会展公司将合作完成国家会展中心项目一至五期的工程开发，其中三期工程资本中心塔现已进入施工阶段。这幢摩天楼内部包括2万 m²的高质量办公、零售空间及阿布扎比的第一座凯悦酒店，建成后将成为国家会展中心开发项目中的标志性建筑，也构成整个阿布扎比城市天际线的视觉焦点。

资本中心塔在设计和建造中的诸多与众不同之处之盛名远扬，尤其是充满诗意和动感的建筑形态给人强烈的视觉冲击。但正如同RMJM结构工程总监Greg Lau指出的那样："当形态呈曲面上升时，建筑将持续受到侧面荷载压力，这对保持建筑结构的稳定性是一种挑战。"塔楼的核心由预应力核心筒及其悬臂部分构成。标准层面积为每层1 000 m²，塔楼平面从核心筒的一侧向另一侧逐步移动，在距顶部4层处完成变形。塔楼18层侧向延伸出一片倾斜面并向下逐渐延展构成遮盖历史大看台的华盖。在19层，一个室外游泳池设置于悬挑结构之上，使凯悦酒店入住者充分享受绝佳的海景。

资本中心塔内部斜肋构架的非对称对角结构能支撑塔楼的外部荷载，这座摩天楼成为"世界上迄今为止第一座使用预应力核心筒的建筑"。RMJM在此项目中引进名为"Cardinal C240"的新玻璃幕墙体系，它的反眩光特性和卓越透明性能保证塔楼内大部分区域白天用可利用自然光照明。双镀银涂层能确保此产品达到眩光最小化、光线传递最大化的效果。

考虑到能源保护是海湾地区的首要任务，塔楼中凯悦酒店部分的楼层都使用双层玻璃幕墙体系，既有效调节室内气候，又为人们无遮挡地俯瞰城市景色创造条件。从空调中排出的气流沿管道在间距为400 mm的双层表皮间流通，形成能降低外表皮吸热量及缓解制冷荷载的缓冲区域。

RMJM has worked collaboratively with ADNEC to deliver Phases 1 – 5 of this important project. Following on from the success of these two phases, Phase 3 the Capital Gate feature tower is now in construction. At 35 storeys high, it will offer over 20,000 m² of high quality offices and retail units as well as Abu Dhabi's first Hyatt hotel. Standing at over 160m tall, the building has been designed to be the landmark statement for the whole of the ADNEC exhibition development. The feature tower acts as a dramatic termination to the organic free-form stainless steel mesh canopy that forms the Grandstand roof and provides a visual anchor for the ADNEC Capital Centre Development as a whole within the Abu Dhabi City Skyline.

Distinctive elements in design and construction chalk up the Tower's claim to fame. Poetic manipulation of form carries intense rhetorical power, but as RMJM's Director of Structural Engineering, Greg Lau points out, "when a building form undulates, twisting and turning away from its centre of gravity, there is a sustained lateral load on the building – a challenge to provide sustained structural stability." The Tower's central axis is its pre-cambered vertical support and its cantilever projection. The average floor area is about 1000 m² per floor. Floors shift gradually from one side of the central core to another, completing the transition on the upper-most four floors. It sustains a lateral slant on the 18th floor and rolls out the Grandstand canopy. An external swimming pool with a magnificent oceanic view overlays the cantilever projection on the 19th floor to indulge Hyatt's visitors.

An asymmetric diagonal structure, a "diagrid", supports the tilting external load of Capital Gate. RMJM's project architect, Tony Archibold, believes that "this is the first building in the world to use a pre-cambered core". RMJM has introduced a new glazing system "Cardinal C240", chosen for its anti-glare properties and excellent transparency so that most of the building can be lit by natural daylight. The product has two silver coatings which minimizes glare and maximizes light transmission.

Energy conservation is a priority in the Gulf. The Hyatt hotel floors are in fact enveloped in a double-skin of glass that enables greater efficiency in controlling the internal climate while providing clear unobstructed views over the cityscape. Exhaust air from the air-conditioned rooms is channeled and circulated within 400mm gap within the double-skin to create a buffer-zone providing thus reducing heat absorption through the external facade and so decreasing the buildings cooling loads.

1 首层平面图
2 建筑整体效果

3 临水立面
4 入口空间
5 剖面图
6 100 m 处标准层平面
7 5—105 层平面
8 26—126 层平面

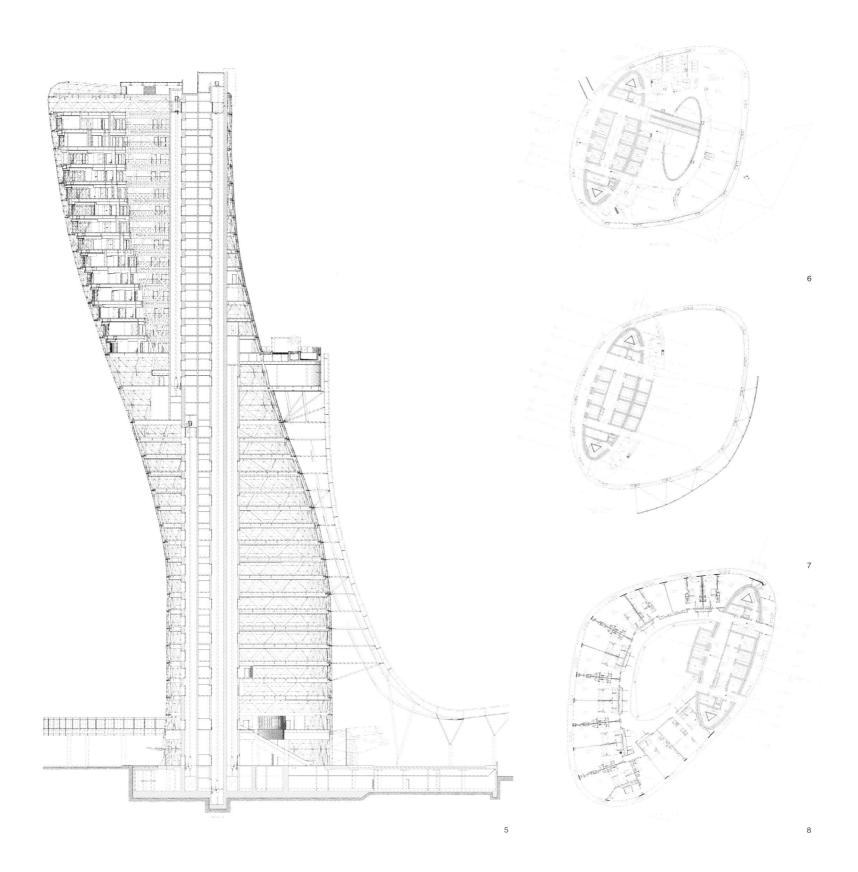

ST. PETERBURGH OKHTA TOWER & CENTRE, RUSSIA
俄罗斯圣彼得堡奥克塔摩天楼

RMJM建筑设计集团 | RMJM Architects

项目业主：俄罗斯天然气公司和圣彼得堡城市管理委员会（Gazprom and St Petersburg City Administration）
项目规模：32.6 万 m^2
建设高度：77 层 /396 m
项目造价：$2.4 billion
项目策划：Office and Amenities
项目设计总监：Tony Kettle（集团设计总监）
设计团队（参与此项目的其他 RMJM 设计人员）：Roger Whiteman, Philip Nikandrov, Clifford Francis, Charles Phu, Jim Patterson, Kirill Zavrajine, Andrey Zolotarov, Frank Theyssen, Ray Bryant, Anthony Tappe, Matthew Johnston
技术顾问：Jonathan Zane, Winslow Kosior
其他项目顾问：Glo——视觉效果小组，Newtecnic——立面构造设计，Whitbybird——结构工程小组，Battle McCarthy——建筑设备工程，Savant——工料计量小组

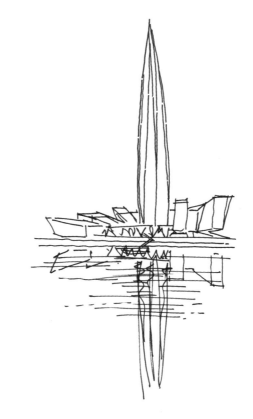

奥克塔摩天楼是RMJM在2006年12月接受俄罗斯天然气公司Gazprom委托承担的设计任务。这座高396 m的螺旋状玻璃摩天楼，位于城市最主要水道涅瓦河畔。

摩天楼坐落于圣彼得堡距历史中心城区5 km外的工业区内，设计灵感呼应于城市历史，效仿市内传统建筑尖顶，好似从城市水平肌理中破茧而出的细针。摩天楼形态创意还与一度耸立在基地中的堡垒形式和水能概念有关。这座城市中新的尖顶建筑由五部分组成，每部分均沿各自的轴线旋转上升，体量也随之减小，直插云霄。同时这种渐变形式也是对水的变化性的一种诠释，通过光的折射和反射幻化出千变万化的外观效果。摩天楼顶层设有公共观景长廊，而在建筑底部，办公区与博物馆、图书馆、健身中心和公共展示大厅一起组成一种校园式空间。

相对于室外温度低至零下30℃的环境，摩天楼的环境策略基于一种"毛皮大衣"的概念，其外表皮由双层玻璃幕墙组成，而中庭作为缓冲区能在一年中的不同时期为建筑提供隔热、保温和自然通风。摩天楼的五角星形平面能最大程度将日光引入室内，既能使办公人员从室内欣赏美丽的城市景色，又不会因表皮暴露在室外而产生失热情况。面积可观的公共交流空间和绿化休憩区域遍布每个楼层，使人们不用耗费时间和体力就能直接进入休闲区域。

Beating off 5 other internationally-renowned architects for the commission, RMJM was appointed in December 2006 to design the Okhta Tower and Centre for client Gazprom. RMJM's winning design features a 396 metre twisting, glass needle which echoes the spires of St Petersburg.

Located 5km from the historic centre of the city in an industrial zone, the inspiration for the RMJM design came from St Petersburg itself, which has a rich architectural tradition of spires breaking the city's horizontal grain. The tower form was also inspired by the fortress that once existed on the site as well as the concept of energy in water. The new spire comprises five elements which gently rotate on their own axis as the building rises and tapers to delicately touch the sky. The site is located on the city's main waterway, the River Neva, with the form of the building deriving its shape from the changing nature of water, ever changing light, reflections and refraction. The Okhta Tower will includes a public viewing gallery at the topmost point and around the base there will be a campus of office buildings together with a museum, a library, sports facilities, and an exhibition gallery which will all be open to the public. In a country where temperatures dip to

1 整体效果

minus 30 degrees, RMJM, in partnership with leading environmental engineers Battle Mc Carthy, designed an innovative energy solution where the need for heating is minimal in order to reduce its environmental impact. The tower's environmental strategy is based on what is being described as a 'fur coat' concept. The external envelope of the tower comprises of two double glazed glass skins with an atrium between the inner and outer walls. The atrium acts as a buffer zone providing both thermal insulation and natural ventilation at different times of the year. The pentagram design of the tower maximises access to daylight and allows for spectacular views for the internal offices without losing heat due to exposed surface area in comparison to other structures. The design allows for a generous number of social spaces and green 'breakout' zones spread out along the floors. These enable office workers to access leisure areas without wasting valuable time and energy using an elevator to reach ground level.

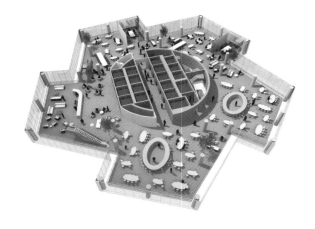

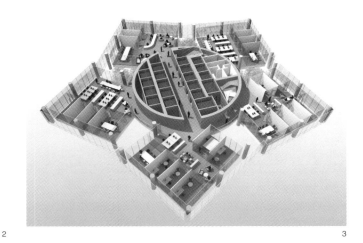

2 十层平面布局模型示意
3 三十四层平面布局模型示意
4 顶层空间效果
5 中庭空间效果
6 通透的表皮

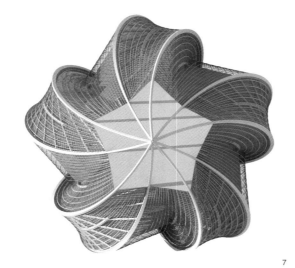

7 顶部形态模型
8 中间层模型
9 底层模型
10 形态模型

11 临水立面
12 夜景效果

ATASEHIR VARYAP PROJECT, TURKEY
土耳其 Atasehir Varyap 项目

RMJM建筑设计集团 | RMJM Architects

项目业主：Varyap 开发 & 建筑集团
项目规模：34 万 m²
建设高度 / 层数：54 层
完工日期：2011 年
项目造价：$500 million
建筑设计：RMJM
当地合作建筑设计团队：Dome
景观设计：Trafo

机械暖通管道设计（MEP）：Happold
结构设计：Buro Happold
当地合作建筑工程设计团队：Yapi Teknik
RMJM 设计小组：Meredith Bostwick, John Brewer, Metin Celik, Alice Wen Chang, Seong Choi, Sang Hyun Chung, Phil Gray, Chris Jones, Winslow Kosior, Kelly Lee, Sung Won Lee, Jing Lu（吕晶）, Cristian Mare, Itir Saran, Peter Schubert, Jonathan Zane
数字化图像创作小组：Ivor Ip（叶思衍）, Andrew Maxwell, Halley Tsai

　　Atasehir Varyap项目所在的基地拥有俯瞰从西Bosphorus海峡到南部Princes和马尔马拉海的全景视野，并将脚下的伊斯坦布尔老城揽入怀中。这个总面积为34万m²的混合功能开发项目主要包括5栋居住塔楼、办公楼和会议设施，同时还配有公共的休闲场地和景观化设计的停车设施。

　　RMJM设计小组将Atasehir Varyap项目构想成兼具创新性和有机性的标志性综合体：创新性体现在规划和可持续性的实践方法上，有机性体现在与伊斯坦布尔城市风光相融合的一体化美中。该项目位于城市规划中的金融区内，并与几处已经建成的住宅开发区域毗邻。这些循规蹈矩地遵循点式布局的高层住宅形制，虽然为开发商提供了简单的开发样式，但却经常以牺牲独特视觉景观为代价，并伴随刻板的定价配置。Atasehir Varyap的设计另辟蹊径，以多样化的方式调配住宅单位的价值结构：1 700套住宅单元被归为15种不同的户型（从一居室到五居室）；同时，在建筑底部顺应结构走势形成的露天阳台为这些住宅单元提供了附加值；室外露台则通过住宅层向室外空间的开放，减少了建筑外表皮的费用及能源消耗。

　　为了更好地利用基地得天独厚的地形优势，形成卓越的观景条件，Atasehir Varyap项目的设计理念旨在平衡观景最大化和热辐射吸收最小化。设计师通过对建筑朝向和形体的推敲，结合景观设计达至两者间的平衡，雨水收集点和废水处理设施优化了用水需求并减少了能源消耗，有机布置的可见的风轮机技与建筑立面相映成趣。大厦外部的蓄水池提升了室外景观的空间质量。

　　在整个设计过程中，Atasehir Varyap设计团队力求融合周边环境的美学要素，并深受伊斯坦布尔独特的人文环境启发，将城市的一些要素引入到综合体的设计概念中，例如项目的总平面形制就是从土耳其传统的几何图样中抽象得到的。建筑师从蓝色清真寺的圆顶、尖塔到Le Corbusier为伊斯坦布尔描画的城市剪影中汲取灵感，使54层高的塔楼犹如从地表破茧而出，直至消失于天际。Atasehir Varyap建筑外表皮的色彩从地面到天空由赤桃色转换到天蓝，这种渐进性的色彩构成源自土耳其室内清真寺的象蓝色瓦片和海峡日出的玫瑰色。

　　Atasehir Varyap项目为商业住宅的可持续设计制定了一个新的业界标准，RMJM设计团队在今后的设计中，将秉持他们将当地文脉和全球化战略相结合的设计理念，延续设计精品建筑的优良传统。

　　In a location that enjoys panoramic views stretching from the Bosphorus Strait in the west to the Princes' Islands and the Sea of Marmara to the south, the Atasehir Varyap project embraces the city of Istanbul from which it arises. The project presents a 340,000 m² mixed-use development including five tower elements housing residential units, offices and conference facilities with landscaped public areas and parking facilities.

　　The RMJM design team envisioned Atasehir Varyap as an iconic complex that would be both innovative and organic – innovative in its approach to planning and sustainability practices; organic in its aesthetic integration into Istanbul's cityscape. Located in Istanbul's proposed financial district, the site is neighboured by several established

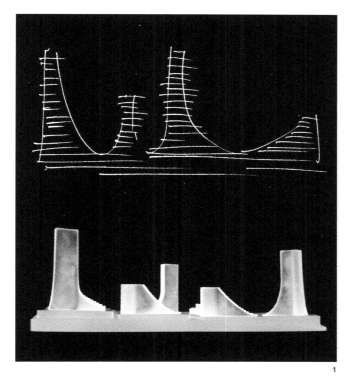

residential developments. The point block organisation typical of such complexes provides a simple solution for developers, yet too often sacrifices unique views and perspectives and result in a rigid pricing configuration. In contrast, the Atasehir Varyap design diversifies the value structure of the residential units; the 1,700 residential flats are divided amongst 15 different flat types in sizes ranging from studio to 5-bedroom format, while below-grade construction has yielded several additional residential units possessing value-increasing open-air terraces. The reorganisation of residential layouts reduces the cost of envelope construction as well as energy consumption by opening up residential levels to park areas by way of the units' outdoor terraces.

To better capitalise upon the topological advantages of its belvedere situation, the design concept for Atasehir Varyap was governed by the need to mediate between the maximisation of panoramic views and the minimisation of solar heat gain. Designers struck this balance through manipulation of the orientation and massing of the buildings, as well as through landscape design and green innovation. Rainwater collection sites and grey water processing facilities optimise water usage and reduce energy consumption while strategically-located visible wind turbine technology, integrated into the building envelope, and

1 概念草图及体量研究模型
2 总体效果

cooling water pools enhance the external landscape.

Throughout, the Atasehir Varyap project preserves a sensibility for the aesthetics of its environment; inspired by the unique context and culture of Istanbul, the development incorporates the city into the complex's design, exemplified in the abstraction of Turkish geometry of the master plan. Citing creative references ranging from the domes and minarets of the Blue Mosque to the long lateral lines of Le Corbusier's sketches of the Istanbul, the buildings seem almost to emerge from the landscape and dematerialise into the sky, taking flight in the form of the 54-story tower. The Atasehir Varyap exteriors are constructed from a spectral colour range which transforms from terra cotta to blue hues in a progression evocative of this rise from the earth to the heavens. The palette of this gradation is derived from sources as varied as the tiles of the Blue Mosque and the rosy hues of the sunrise over the Strait.

The RMJM team continues to create international works distinguished in their site-specific yet universally-accessible design. The Atasehir Varyap project promises to set a new industry standard in sustainable commercial design, continuing in a tradition of excellence in architecture.

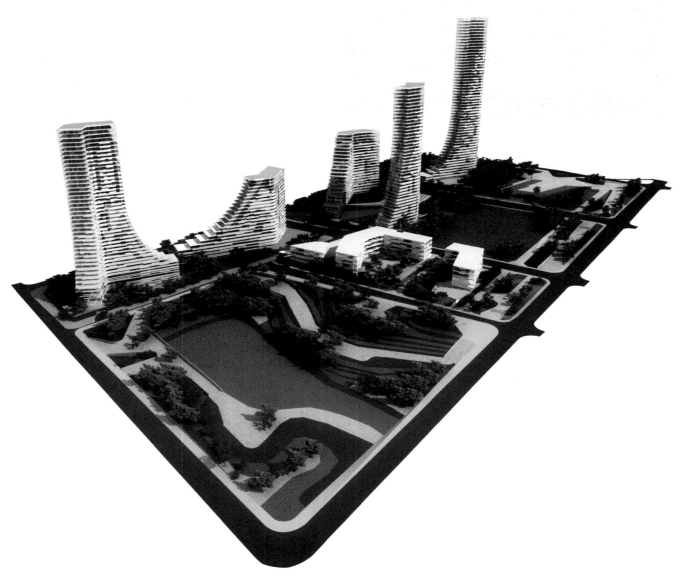

3 鸟瞰

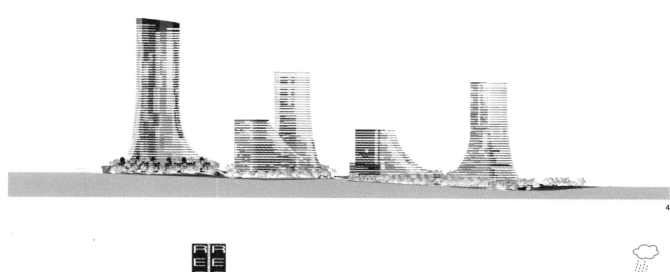

4

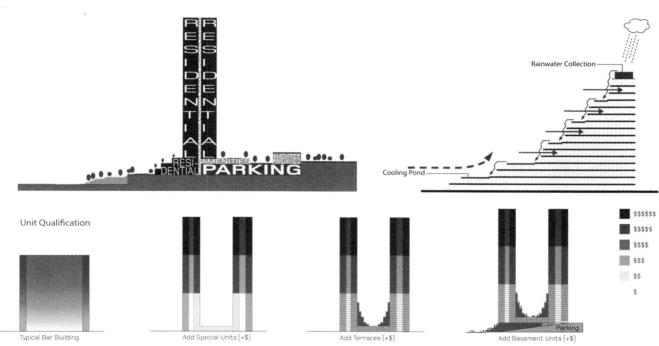

5

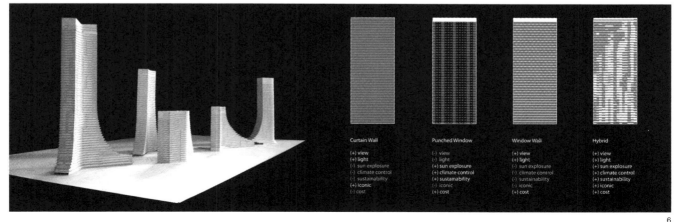

4 立面
5 竖向功能分析
6 表皮及体量构成分析

6

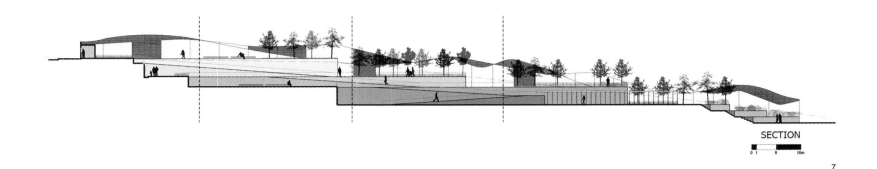

7 SECTION

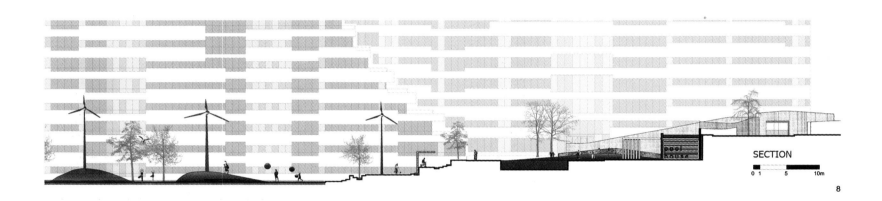

8 SECTION

9

10

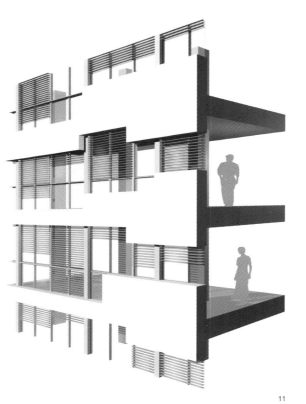

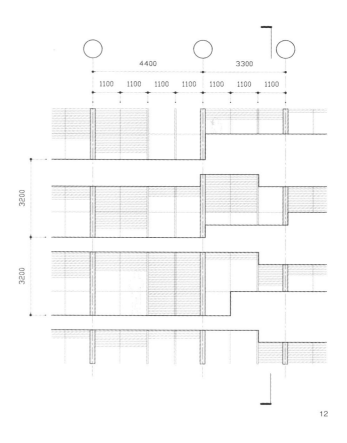

7 景观剖面
8 景观剖面分析
9 景观局部透视
10 室内效果
11 立面剖透视
12 部分立面

11　　　　　　　　　　　　　　　12

NANJING SUNING WEST RIVER CITY PLAZA
南京河西新城苏宁广场

凯达环球 | Aedas Ltd.

建筑设计董事：温子先（Andy Wen）
占地面积：32 996 m²
总建筑面积：35 m²
建筑高度：450 m

苏宁广场共118层，高450 m，为使其能够成为一座标志性建筑，设计师将建筑的象征主义和未来功能完美结合。从总体外观来看，整座建筑并不是突兀地指向天空，而是自然地从地下"生长"出来，究其原因是宽阔的建筑底盘与锥形楼体间的自然过渡。

受东方传统理念"阴"和"阳"的启发以及罗丹著名雕塑《吻》的影响，设计师创造出双塔形式，随着高度而和谐地交叉重叠，并相互交融、相互补充。两座塔楼相互对照并保持自身的独立性，外表皮分别采用石化玻璃和钢化玻璃，象征"女性"和"男性"特征。整座建筑有着复杂的曲线、变化的半径和层次，力图通过多角度变化来改变自身形态，建筑表面在自然环境和人工光线的作用下更突出扭转的效果，在某种程度上也让人联想起翩翩起舞的姿态。建筑底部层叠的立面依次向外伸展，且在竖向高度上形成遮阳功能。

在技术层面，苏宁广场充满了创新性，成为最具可持续发展价值的建筑物。如建筑内部通风由透气双层幕墙提供，能从建筑物底部抽取空气，形成对流，空气也通过风道横向流动。建筑内部安装的涡轮机除减轻建筑物所承受的风压外，还能用来产生"清洁"电能。大厦顶部的一组太阳能电池使整座建筑不再依靠电网供电。

大厦总楼面积为35 m²，其中5~7层设置令人叹为观止的下沉广场及公共空间，且与地铁直接相通，供商业和零售功能使用。办公区设在18~23层，其上是28层的宾馆和服务式公寓，最高的23层功能介于办公区和服务式公寓之间。参观者站在整个结构顶部的大厅可俯瞰南京城市全景。

At 450 metres and 118 floors, Suning Plaza's height firmly establishes it as a city landmark. To also make it an icon, the architects endowed it with a form that represents a seamless marriage of symbolism and forward-looking function. By its general appearance, the building does not seem to violently stab upwards but rather 'grow' from the earth like a natural phenomenon, an impression largely created by the transition between the severe rake of its broad podium and the subtle taper of the tower component.

The most arresting characteristic of the tower, however, is that it is conceptually not one structure, but an intertwining of two. Inspired by the traditional Eastern notion of 'yin' and 'yang' as well as the famously intimate Rodin sculpture *The Kiss*, the architects created a pair of tower forms that both overlap and twist in concert as they rise. They are sensual, complementary and yet distinct, with contrasting skins of stone-and-glass and steel-and-glass respectively symbolising 'feminine' and 'masculine' characteristics. With its sophisticated curves, radii and layers, the building seems to transform itself with every change of perspective, its surfacing accentuating the effect as it reacts to the play of natural and artificial light. The overlap between the two forms is in itself a fascinating aspect of Suning Plaza. Viewed from certain angles it brings to mind the side-slit of a 'qipao' dress, making the building form somewhat reminiscent of a dancing couple.

1 建筑入口效果

HIGH-RISE BUILDING | DESIGN WORKS

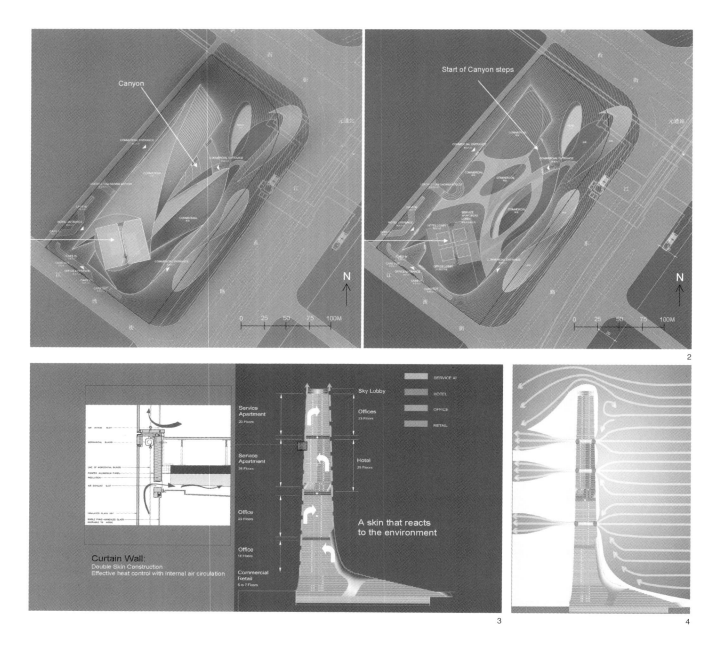

The outward flare of the overlap also forms a natural entranceway into the podium, and creates an integral sun-shade along its entire height.

Technically, Suning Plaza's feast of innovations will make it one of the most sustainable buildings ever conceived. Much of the ventilation for its interior volumes, for example, is provided by a 'breathing' double-skin curtain wall which draws a convective current of air up from the base of the building. Air also flows horizontally through 'wind tunnels' placed at three widely separated levels of the structure. As well as relieving wind pressure against the structure, turbines installed within these features can also be used to generate 'clean' electricity. An array of photovoltaic cells on the roof of Suning Plaza will further bolster the building's independence from the electricity grid.

Inside, Suning Plaza offers a maximum GFA of 350,000 square metres. The five-to seven-floor podium – which includes a stunning sunken plaza/communal space as well as direct links to the underground metro – is devoted to commercial/retail use. Immediately above are 18- and 23-floor segments of office space, followed by 28 floors of hotel and serviced apartments, and a final 23-floor segment split between offices and serviced apartments. Crowning the whole structure is a 'sky lobby' offering panoramic views over the Nanjing cityscape.

2 建筑平面功能分析
3 立面分析
4 表皮通风分析
4 建筑立面效果
5 临街效果

3

5

EMPIRE TOWE, UNITED ARAB EMIRATES
阿拉伯联合酋长国帝国大厦

凯达环球 | Aedas Ltd.

建筑设计总监：安德宝（Andrew Bromberg）
结构、机电工程师：Whitby & Bird
设计时间：2006 年

  2006年，凯达接受帝国控股集团的委托，在阿联酋阿布扎比Al Sorouh海岸一处面积为7 013 m²的黄金地段设计一栋60层的豪华住宅楼。项目基地三面毗邻主要街道，属大规模规划范围内的主体部分，项目周边可能布满标志性建筑。因此，对该客户而言建筑的标志性是至关重要的，设计既要在林立的楼群中塑造自身的独特风格，但又不能"太出格"沦为建筑行业的负面案例，它应轻而易举地给人留下深刻印象。

  凯达以此为出发点，分析基地的劣势与优势——优势是东北向可鸟瞰大海，西南向可欣赏园林景观；劣势则是其紧邻大型商务楼。因而帝国大厦采用既不流俗又不出格的外观形态来应对这一挑战。

  帝国大厦引人瞩目的造型设计主旨是最大限度地保持与街道环境的和谐，又与周边商业楼截然不同。因此，楼层在底部向四面伸展开，如同树根一样，塔身部分倾斜直上，在达到一定的斜度后回转。建筑侧翼九个锋利的"刀片"结构与建筑物正面的垂直表面及柔和轮廓形成鲜明对比，且其中的六个"刀片"从地面直达238 m高空。除了无可争议的视觉效果外，大厦的复合形式也带来了实际效能，使邻近商业楼和街区外的海面景观通道得以扩大。同时，"刀片"结构最大限度地扩大每个单元的正面空间并拓宽视野。最终大楼单元中70%可看到大海，其余30%可尽览附近公园景致。

  帝国大厦的另一特点是南面朝向的阳台，除使居住其中的人能以更宽广的视野、更近距离地观赏周边景观，这种南向阳台的一个实际用途就是为下面房间遮阳。与此相反，大楼北面也朝向阳光建造，以便最大限度地取得自然采光。同时，"刀片"结构外部覆盖隔热玻璃幕墙，能高效散热并进行精细调节，以适应当地气候条件，此外，幕墙与众不同的色泽也增加了大厦的美感。

  真正造就大厦非凡外形的主要因素是其创新和异常高效的结构框架。设计师没有简单地将剪力墙放在沿核心带内侧的传统位置，而是放到走廊外缘。这一措施拓宽了结构基础，减少核心带与表面间的距离及整体结构深度，最终减化繁多的结构构件。

  帝国大厦独特的形态所带来的另一意想不到的效果是：居住单元尽管随大楼的倾斜而同步水平"移动"，但其尺寸和布局大多能实现标准化。60层建筑所有楼板面积为9.54万 m²，建筑物核心带集中并垂直地堆建，最大限度提供实用性和多功能性。东西两侧采用模块化单元，随单元位置逐渐转移，形成与众不同的"刀片"结构。借助59层和60层间面积较小的楼板和核心筒，形成独特的复式单元。

  Unsurprisingly given its striking originality, this 60-storey luxury residential building evolved from an unusually challenging brief. In 2006, The Empire Holdings approached Aedas, the well-known global architecture practice, to work on a design for a prime 7,013 square-metre plot near the coast at Al Sorouh Abu Dhabi in the UAE. Bordered by three major streets, the site is part of a larger masterplan that places a clutter of potentially iconic buildings within close visual range of Empire Tower. It was vitally important to the client, therefore, that the design 'cut through' with a singular visual presence, but not 'push the envelope' so far as to emerge as a mere architectural caricature. The Empire Tower would have to leave its imprint without trying too hard.

  Led by design director Andrew Bromberg at Aedas, the team began their design effort by intensively scrutinising the site and discovering the challenges and advantages it posed. Among the latter are ocean views to the northeast and park views to the southwest. Among the former was the immediate presence to one side of the site of a large commercial tower. The Empire Tower evolved its response to both through an outstanding balance of form and alignment.

  The tower's eye-catching form grew from the desire to maximise its street-level presence whilst establishing an identity apart from its commercial neighbour. Thus, its form splays widely at the base, like the root system of a tree, inclining away from the street as it progresses

1 临街效果

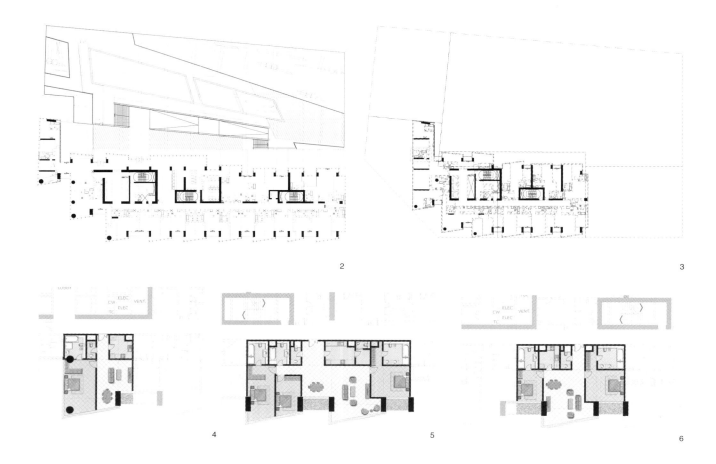

upwards before transitioning to a moderate slant in the opposite direction. Contrasting with the sheer surfacing and soft contours of the building's 'face' are nine sharp-edged 'blade' structures punctuating its flanks, six of which rise from ground level all the way to the tower's 238-metre total height. Apart from its undeniable visual impact, the tower's complex form brings the practical benefit of enlarging the view corridor past the neighbouring commercial building to the sea one block away. The 'blades', meanwhile, serve to maximise individual units' frontage, and hence their views. Overall, 70 percent of the tower's units boast sea views, with the remaining 30 percent enjoying superb views of the nearby park.

Adding yet further characteristic detail to Empire Tower are its south-facing balconies. Besides allowing residents' an even wider, more intimate view of their surroundings, these features serve a practical purpose by shading the apartments below. The southern facade as a whole has been angled to avoid direct solar gain. Conversely, the tower's north face is oriented to maximise natural light. The blades, meanwhile, are externally clad with an insulated glass curtain wall. Thermally efficient and fine-tuned to match local climatic conditions, the glass also makes a contribution to the building's aesthetic imprint thanks to its distinctive tint and reflectivity.

Literally underpinning this unusual form is an innovative – and extraordinarily efficient – structural scheme. Rather than placing the building's shear walls in their traditional location along the inner side of the core, the architects behind Empire Tower pushed them to the outer edge of the corridor. At a stroke, the effect was to widen the structural base, reduce the distance between core and facade as well as the overall structural depth, and ultimately, reduce the mass of the structural members.

Unexpectedly given its external contours, the residential units of Empire Tower are largely standardised in size and layout, though they horizontally 'shift' in step with the building's inclination. All 60 of the floor plates that give the development its 95,411 square-metre GFA are 'non-typical'. By contrast, the building core is centralised and vertically stacked, maximising ease of construction and functionality. Modular units were employed on the eastern and western sides, their individual positions progressively shifting to form the structure's distinctive 'blades'. Smaller floor plates and a reduced core area on levels 59 and 60 allowed for the creation of a unique duplex unit.

2 标准层平面
3 三十层平面
4 户型1的空间格局
5 户型2的空间格局
6 户型3的空间格局
7 剖面
8 剖面

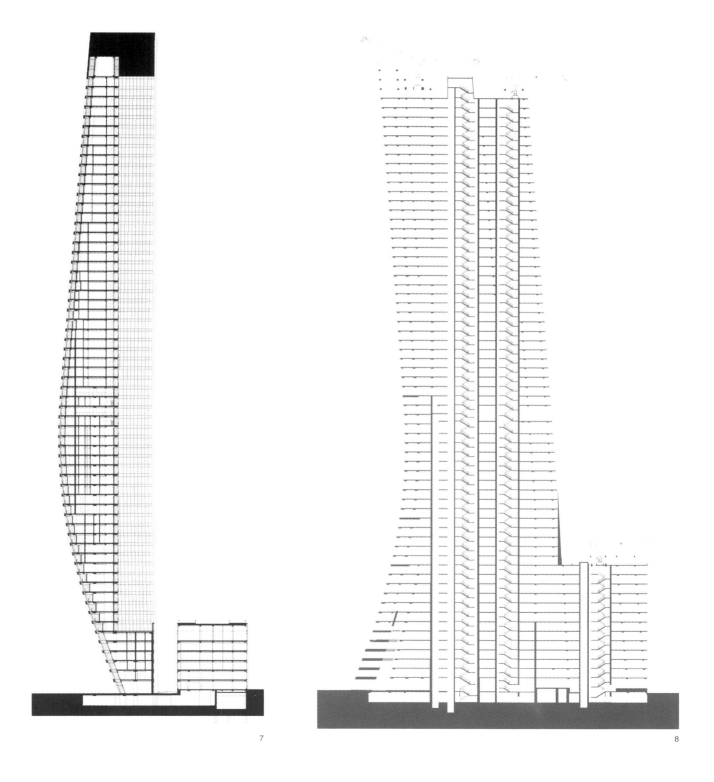

7

8

DUBAI PENTOMINIUM TOWER, UNITED ARAB EMIRATES
阿拉伯联合酋长国迪拜 Pentominium 大楼

凯达环球 | Aedas Ltd.

建筑设计总监：安德宝（Andrew Bromberg）
结构工程师：Hyder
设计时间：2006年

Pentominium Tower位于迪拜Marina港，高516 m，是世界最高的、也是最具特色的住宅楼之一，它采用细长的轻型结构，通过诸多令人着迷的细部和特征打造全新的现代豪华生活方式。Pentominium的精髓在于豪厦与阁楼的完美结合。这种设计允许建筑的一层为超大的空间单元，内部包含豪宅应有的超级便利设施和豪华配置。设计的另一重要环节是对两种制约因素的克服，一是适应高度密集的邻里结构，二是应对迪拜固有的环境压力。

基于这些难题，建筑主体两侧形态各异，围绕共同的核心区域构筑出众的外形与功能。建筑朝南的一面采用简单外向的突出结构，与楼体等高。包含一系列阳台和垂直分层的玻璃，用以将吸收的太阳光热量保持在可接受的范围。随着建筑物不断向上攀升，这一玻璃层逐渐变宽，形成有机防护层，保护阳台免受高空强大风力的破坏。

与此相反，建筑物朝北的一面呈块状结构，公寓和空中花园错落交替，使具有吸引力且又功能丰富的公共与半私密空间得以共存，Pentominium也因此能在拥挤的环境中"自由呼吸"。Pentominium的出众构造还带来另一种出众功能，那就是它在实质上克服细长型超高建筑物常见的"摇摆不稳"的难题。Pentominium之所以能实现适度的"稳固性"，关键在于它采用不规则的垂直结构。这些结构经过精心调整，可打破风流旋涡，避免引发容易招致麻烦的横风风振反应。因此，对于同等高度的细长建筑而言，Pentominium能承受前所未有的动力负载和加速度。此外，支撑建筑物的大量桩柱也成就了Pentominium不同寻常的稳固性。为达到最大强度，每个桩柱的直径为900 mm～1500 mm，地下嵌入42 m深。桩柱系统的异常深度使Pentominium成为中东地区最能"深入地下"的建筑物，也构筑了世界上最高的住宅楼。

Pentominium的独特布局和结构不仅表现在技术层面上，同时也是营造内在豪华设施的基础。由于住宅单元上面安装的跃层楼板，使位于6层的俱乐部会所具有异常的层高。101层集中布置多个高层娱乐会所设施，包括空中酒吧、商业中心、雪茄吧和令人赞叹的露天观景台等。创新设计还为Pentominium住户带来无与伦比的便利。如速度极快的电梯系统令人印象深刻，其他同类建筑物通常采用门廊处转乘电梯的方法，而Pentominium则采用"直梯"方法，使其直接运行500 m，这是史无前例的。因此，用户只需等待预设的40秒，甚至更短时间就可乘坐电梯，这一成就对高达516 m的建筑而言，真是令人称奇。Pentominium以其优雅外形和创新的结构设计，必将为迪拜带来醒目的空中景观。一面1:14.3的长细比（即宽度与高度比）及另一面1:12.28的比例，最终造就这座富有动感、形态轻盈的516 m高的豪华城市景观，它在与周围环境浑然一体的同时，又以独特的标志性气势震撼人心。

The 516-metre Pentominium Tower at Dubai Marina is one of the tallest residential buildings in the world. It is also one of the most distinctive, with a slender, rapier-like form embodying a number of fascinating details and features to create a new paradigm in modern luxury living. The essence of Pentominium is its conceptual fusion of the condominium and the penthouse. This allows for a single ultra-spacious unit per floor, containing all the superb amenities and premium luxury features expected in a property of this status. Another crucial aspect of the design brief was the need to respond to the high density and close proximity of neighbouring structures, and to the extreme environmental factors inherent to Dubai.

With these challenges in mind, a team of Aedas architects led by design director Andrew Bromberg evolved Pentominium's general configuration of two different sides centred around a shared core. In an outstanding example of form following function, one side of the structure consists of a simple extrusion that extends to the full height of the tower. This side is primarily southern-oriented and also features a system of balconies and vertically layered glass that keep solar heat intrusion to acceptable levels. As the building ascends, this layer of glass broadens, to form an integral 'shield' that shelters the balconies from the powerful winds usually experienced at higher altitudes.

By contrast, the opposite side of the building takes on a segmented profile, alternating between apartment volumes and voids containing sky-gardens. Adding to their distinction – both aesthetically and technically – the apartment volumes comprise of six five-storey-high pods that cling structurally to the building core. Their alternation with the garden voids allows for an engaging and highly functional mix of communal and/or semi-private spaces, and enables Pentominium to 'breathe' within its dense context. Another way in

1　建筑与环境的融合

2 空中休闲区域
3 空中花园示意
4 建筑体量演化分析
5 建筑临街效果

which Pentominium's outstanding form results in outstanding function is its virtual elimination of the 'swaying' usually experienced in tall, slender buildings. The key to Pentominium's comfortable 'stability' are the irregular vertical profiles found in its external form. These were meticulously fine-tuned to break up vortices that would otherwise generate undesirable cross-wind responses. As a result, Pentominium experiences aerodynamic loads and accelerations that are unprecedentedly moderate for a building of its equivalent height and slenderness. Also contributing to Pentominium's outstanding stability is the extensive system of pilings that underpin the structure. Measuring 900 to 1,500mm in diameter for maximum strength, each pile is embedded to a depth of up to 42 metres below ground level. The exceptional depth of this piling system makes Pentominium the 'deepest' structure in the Middle East as well as the tallest residential building in the world.

The advantages of Pentominium's unique layout and structural scheme are more than technical – they are the basis of the building's inherent luxuriousness. The lower club house level on L6, for example, boasts an exceptionally high ceiling thanks to the staggered floor slabs of the residential units directly above. On Level L101 are Pentominium's extensive upper club house amenities, including a sky lounge, business centre, cigar bar, and a stunning open-air observation deck. Sweeping long-range views of the Dubai skyline are guaranteed by Pentominium's record-breaking height and generous external glazing. Innovative engineering also gives Pentominium's occupants an unprecedented level of convenience. The building's high-speed lift system is a particularly impressive example of state-of-the-art design. In contrast to other buildings of this stature which typically employ lobby transfer systems, Pentominium employs a 'direct' lift system with an unprecedented straight-run of 500 metres. Thus, users need wait for a projected 40 seconds or less for lift service – an astonishing achievement for a 516-metre building.

Pentominium's elegant form and innovative architectural design will make a striking visual impact on the Dubai skyline. With its slenderness ratio (width-versus-height) of 1:14.3 on one side and 1:12.28 on the other, the end-result is a dynamic, lightweight 516-metre tower that takes its place naturally within a setting of extruded neighbours, whilst still maintaining a powerful, iconic presence that is uniquely its own.

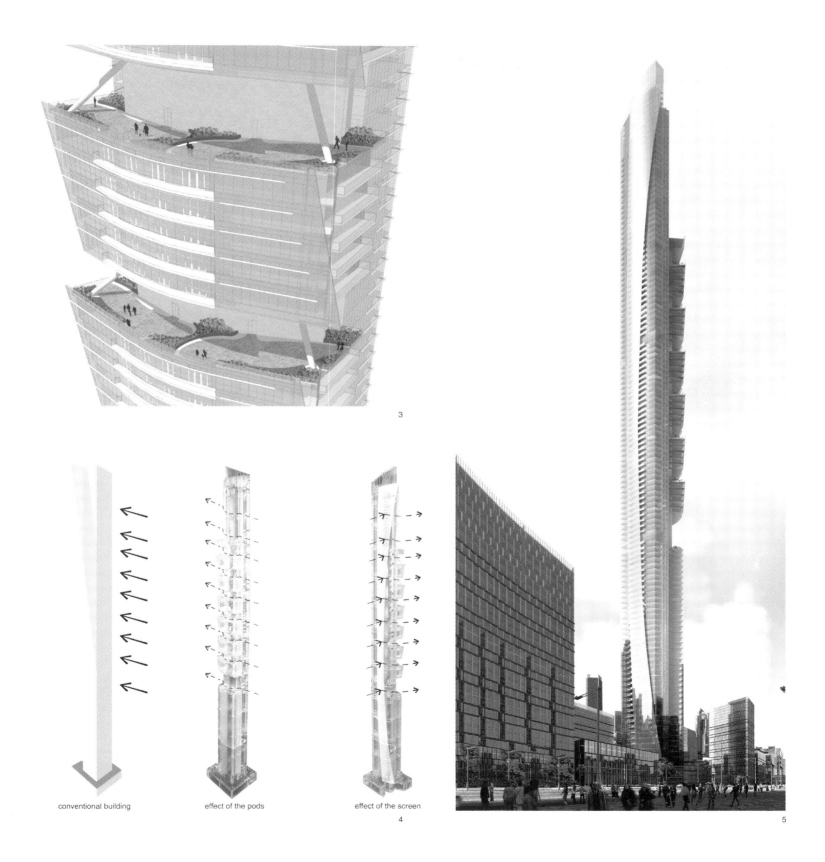

conventional building   effect of the pods   effect of the screen

图书在版编目（ＣＩＰ）数据

高层建筑 / 梅洪元, 朱莹主编. -- 哈尔滨：黑龙江科学技术出版社, 2014.12
（当代建筑创作理论与创新实践系列）
ISBN 978-7-5388-8134-9

Ⅰ. ①高… Ⅱ. ①梅… ②朱… Ⅲ. ①高层建筑 - 建筑设计 - 研究 Ⅳ. ①TU972

中国版本图书馆 CIP 数据核字(2014)第 294546 号

当代建筑创作理论与创新实践系列——高层建筑
DANGDAI JIANZHU CHUANGZUO LILUN YU CHUANGXIN SHIJIAN XILIE——GAOCENG JIANZHU

作　　者	梅洪元　朱　莹
责任编辑	刘丽奇　王　姝
封面设计	赵雪莹　赵天杨
出　　版	黑龙江科学技术出版社
	地址：哈尔滨市南岗区建设街 41 号　邮编：150001
	电话：（0451）53642106　传真：（0451）53642143
	网址：www.lkcbs.cn　www.lkpub.cn
发　　行	全国新华书店
印　　刷	北京顺诚彩色印刷有限公司
开　　本	889 mm×1194 mm　1/12
印　　张	20.75
字　　数	350 千字
版　　次	2014 年 12 月第 1 版　2014 年 12 月第 1 次印刷
书　　号	ISBN 978-7-5388-8134-9/ TU・715
定　　价	198.00 元

【版权所有，请勿翻印、转载】